AN ENVIRONMENTAL HISTORY OF BRITAIN

Frontispiece :
[Arkwright's] Cromford Mill, 1789 by William Day. Industry begins to encroach on Nature, but the artist detects no adverse effects on the environment.

An Environmental History of Britain

since the Industrial Revolution

B W Clapp

Longman
London and New York

Longman Group UK Limited,
Longman House, Burnt Mill,
Harlow, Essex CM20 2JE, England
and Associated Companies throughout the world.

Published in the United States of America
by Longman Publishing, New York

First published 1994

ISBN 0 582 22627 9 CSD
ISBN 0 582 22626 0 PPR

British Library Cataloguing-in-Publication Data

A catalogue record for this book is
available from the British Library

Library of Congress Cataloging in Publication Data

Clapp, B.W. (Brian William)
An environmental history of Britain since the Industrial Revolution /B.W. Clapp.
p. cm.
Includes bibliographical references and index.
ISBN 0-582-22627-9 (CSD). – ISBN 0-582-22626-0 (PPR)
1. Great Britain – Environmental conditions – History. I. Title.
GE 160.G7C57 1994 93-13073
363.7'00941–dc20 CIP

Set by 7 in 10/12pt Baskerville
Produced by Longman Singapore Publishers (Pte) Ltd.
Printed in Singapore

Contents

List of Abbreviations

AAS	*Annual Abstract of Statistics*
BLPES	British Library of Political and Economic Science
CEGB	Central Electricity Generating Board
CPRE	Council for the Protection of Rural England
DoE	Department of the Environment
DSIR	Department of Scientific and Industrial Research
EcHR	*Economic History Review*
GLC	Greater London Council
MRW	*Materials Reclamation Weekly*
NSAS	National Smoke Abatement Society
NSCA	National Society for Clean Air
PP	Parliamentary Papers
PRO	Public Record Office
qu/qq	question(s)
r.c.	Royal Commission
s.c.	Select Committee
SPAB	Society for the Protection of Ancient Buildings
WTW	*Waste Trade World*
v.d.	various dates

Readers should note that the *Municipal yearbook* is published by *The Municipal Journal*. The place of publication for books is London unless otherwise stated.

List of Illustrations

PLATES

MAPS

Acknowledgements

The publishers would like to thank the following for their permission to reproduce illustrative material: Derby City Council, Museum and Art Gallery for '[Arkwright's] Cromford Mill'; Southampton City Art Gallery for 'Haytime in the Cotswolds'; ©Design and Building Services, Sheffield City Council for 'Sheffield, a smoky city'; The Hulton Deutsch Collection Limited for 'A winter's morning in the West End of London, 1924'; The British Library for 'A letter written in Calder water, 1868' and the '*Waste Trade World*: cover'; © Environmental Picture Library / Dave Townend for 'The beginning of the M40 at Uxbridge'; © E P L / Dave Ellison for 'The Stanlow refinery'; © E P L / Martin Bond for 'A slate quarry'; Punch for 'The letter killeth'; Catalyst: The Museum of the Chemical Industry for 'The Kurtz Alkali works, Sutton'; ©Derek Laird / Still Moving Picture Company for 'The two Forth Bridges'; the Controller of Her Majesty's Stationery Office for the map of air pollution in Britain.

Preface

This book is a foray into environmental history, a branch of historical writing not yet widely practised in Britain. The scope of environmental history has not been settled in theory or by long practice, and the author of a book on the subject therefore has an opportunity, and perhaps a duty, to explain what ground he hopes to cover. It would be rash to dogmatise that the chosen territory is the only possible one on which to do battle. But if it serves no other purpose, defining the scope of a book may be a useful defensive measure: it disarms the critic who likes to complain that, if he were writing, this is not the book he would have written. What topics, then, should environmental history cover? Pollution must be included: it destroys some of the amenities of life – clean air, limpid water, wholesome land. Other amenities – natural beauty, old buildings, access to the countryside – also need consideration. The depletion of the country's natural wealth – land, fossil fuels and other minerals – and the ways in which such depletion has been lessened or prevented, are perhaps the most important aspects of environmental history. According to this view, natural resources and the architectural heritage are the proper study of the environmental historian. In the eyes of the economist and the ecologist, man himself is a natural resource. It is perfectly clear that the economic system has wasted man as well as nature and the memorials of times past. Poor education, degrading and unhealthy work, and mean housing have a long history. They did not begin with the industrial revolution nor have they ended yet. These themes have had and will continue to have their historians. But they are distinguishable as a separate area of study that falls within the province of social history. This is a subject much more carefully worked over than the impact of the economy on natural resources and man-made amenities. In surrendering it willingly to the social historian I am sparing readers the threat of a work at least three times as long as the present one.

Fools rush in where polymaths fear to tread. I am by trade and training an economic historian, and economic history is itself a mixed discipline requiring several kinds of expertise. To help in the study of the economy of past times, the ideal economic historian must have some understanding of economics, political history, technology and the science underlying technology. If he ventures into the field of environmental history it will be useful to have a nodding acquaintance with the life- and earth-sciences as well. He may also find himself pronouncing on matters within the province of the architectural historian. Admirable as it may be to breach the fences between academic disciplines, the academic huntsman who careers too wildly may break his neck before he runs down his fox. It is for the reader to decide whether this particular huntsman has had a good or a disastrous day in the field.

Like all authors I am happy to acknowledge my debt to others. Without the willing help of librarians and their staff this book could not have been written. I owe particular thanks to the British Library of Political and Economic Science, and to the newspaper library at Colindale, a division of the British Library. Longman Higher Education and an anonymous adviser made some valuable suggestions. Dr Edmund Newell gave me the benefit of his expert comments on the chapters about atmospheric pollution. My wife returned good for evil: she put up with long periods of silence and then gallantly made neat copy out of an untidy manuscript, at the same time keeping a critical eye open for mistakes and ambiguity. My brother read the proofs, and pointed out (not always in vain) loose expressions, doubtful views and old-fashioned words. I am also grateful for Rodney Fry for drawing the maps. My mother and father taught me the value of thrift, which is in a sense the subject-matter of this book, and like other poor people practised recycling long before it became a popular term or a fashionable virtue.

Overleaf:
'Haytime in the Cotswolds' by James Bateman. Man made this landscape, yet, mistakenly, we do not feel that Nature has been affronted.

CHAPTER ONE

Introduction

THE FIRST CONSERVATIONIST

The origins of the conservation movement are not easy to trace. Malthus, the founder of demography, has some claims to be regarded as the first conscious and celebrated conservationist, though he would not have understood the term itself, coined long after his death. The word 'conservation' in its popular modern sense – the careful husbandry of natural resources and of the meritorious works of man – first came into use in the United States in the early years of this century. It remained little known in Britain until the 1950s. Bertrand Russell, the philosopher and Nobel Laureate for literature, took pains to give the term a broad environmental meaning as if he feared that that usage was unfamiliar to his audience:

> I mean by 'conservation' not only the preservation of ancient monuments and beauty spots, the upkeep of roads and public utilities, and so on. These things are done at present, except in time of war. What I have chiefly in mind is the preservation of the world's natural resources. . . .[1]

Malthus a century-and-a-half earlier had concerned himself not with natural resources in general but with food supplies. In his original 1798 essay on population he detected 'an essential difference between food and wrought commodities, the raw materials of which are in great plenty. A demand for these last will not fail to create them in as great a quantity as they are wanted'. Quite why Malthus held this view is a minor mystery. When he wrote, the raw materials of industry were mostly of organic origin like food itself. The laws that limited food production might be supposed to apply to the production of wool, cotton, leather and timber. If growing numbers threatened to press upon the means of subsistence they would surely press upon the means of housing and clothing also. Malthus apparently came to realise this for, in the last edition of the *Essay* published in his lifetime, he expressed himself rather more cautiously: 'The powers of the earth in the production of food have narrower limits than the skill and taste of mankind in giving value to raw materials.'[2] These are not the words of a man seriously anxious about the extent of natural resources in general, and it is upon his demographic theories that Malthus' reputation as a conservationist will have to rest. Though by no means the first political or economic commentator to notice the possible dangers of population growth – he had been preceded by Machiavelli, Sir James Steuart, Adam Smith and Robert Wallace – Malthus was the one writer in this group who drew out the full implications of the 'principle of population' and who succeeded in attracting wide and often hostile attention to his views.[3]

Malthus shifted his ground during the thirty years in which he wrote and rewrote his *Essay on the principle of population*, but he never abandoned the key point of his theory – that, if unchecked, population would grow indefinitely, perhaps doubling every generation. Food supply on the other hand was unlikely to grow so fast owing to shortage of land and inadequate agricultural techniques. His tone changed from the confident exaggeration and brutal expressions that slipped from his pen and gave so much offence in 1798 and 1803, to the liberal and humane concern for the labouring classes that obviously preoccupied him in the later editions of the work. And his early deep pessimism that little could be done to improve the lot of the poor gave way to a moderate hope that moral restraint combined with wise laws might improve living standards, lessen inequality, raise the status of women and promote freedom for the labouring poor. In all this Malthus concerned himself with what J S Mill was to call 'the niggardliness of nature', the prospective shortage of food if demand continued to grow. Awareness of the dangers of growth marks Malthus out as a conservationist.

GROWTH RATES

The key idea in the Malthusian system was the mathematical proposition that 'no quantity can increase by compound interest and remain within finite limits'. A practical example of the working of this theorem may be helpful. In parts of Asia, Africa, and South America the population has increased since the Second World War at the rate of three per cent per annum compound. At this rate population doubles in 24 years, doubles again in 48 years . . . and therefore in 1000 years doubles more than 40 times! In other words, if such growth were sustainable, population would have increased 2^{40} or a million, million times. Malthus had thought that such rapid growth could occur only in the most favourable conditions, which he detected in the United States. Elsewhere, he thought, slower growth was to be expected, but that would only postpone, not prevent, the day of reckoning when population became impossibly large and a finite world could no longer sustain it. What is true of population is also true of natural resources, mineral as well as organic; it applies as much to coal and iron as to wheat and timber. It is physically impossible to increase the production of these or any other commodities at compound interest for ever and ever.

Long before the idea of compound interest was applied to growth of population and the supply of raw materials, lenders had applied it to money. The Babylonians in 1600 BC were charging compound interest on loans – but in a world that regarded all interest charges as morally doubtful, interest on interest was decidedly unpopular. In the Middle Ages, if creditors charged compound interest, they took care to conceal the fact, and few examples have been discovered by historians. These scruples were being overcome in the sixteenth century, and the first printed tables of compound interest were published at Lyons in 1558 by Jean Trenchant. The first English tables were published by Richard Witt in 1613, and a second set by William Webster appeared in 1629.* Refinements in the use of compound interest allowed the true costs and benefits of decisions about savings and investment to be calculated. Insurance, land management, forestry, and mining could all gain from careful appraisal of true economic costs. The

* According to Walford, *The insurance cyclopaedia* (1871) I, sub Annuities on lives, a first edition of Webster appeared in 1620; the British Library does not have a copy, nor is it recorded in Wing, *Short-title catalogue.*

astronomer Edmund Halley (1656–1742) and the Huguenot mathematician Abraham de Moivre (1667–1754) provided the theory,[4] and insurance offices, landowners, mining engineers, and foresters sooner or later began to apply it – Dutch insurers in the seventeenth century, some land stewards in the eighteenth, mining engineers and foresters in the nineteenth. Economists and economic historians were slow to use compound interest as a measure of growth rates. Hoffmann used them in his prewar study of British industry, but it was not until the 1950s and 1960s that it became second nature among British economists and economic historians to measure growth rates in this way.[5] For advocates of growth – and for their critics – the idea of compound interest is now an important analytical tool. If compound interest did not exist it would not be necessary to invent it before discussing environmental questions, but its existence does permit a more precise account of growth and the dangers of growth.

ECOLOGY

If compound interest is the weapon that mathematics has given to the environmental movement, biology has supplied one even more powerful in the comparatively new study of ecology. The American writer H D Thoreau coined the term ecology in 1858, without attaching a very precise meaning to it. It was Ernst Haeckel, a German biologist, who gave the word scientific currency with the meaning 'all the various relations of animals and plants to one another and to the outer world.' A more homely definition, and one in keeping with the word's derivation, would be 'Nature's housekeeping'. The idea preceded the term. In 1859 Charles Darwin gave a neat illustration of ecological forces at work when he demonstrated the connection between red clover and cats: the bumble-bee pollinates red clover, but is preyed upon by the vole; the cat, which eats the vole, is therefore a friend to bumble-bees and red clover. Other biologists playfully extended the analysis: red clover fed the cattle that provided the beef on which England's greatness depended (Carl Vogt). T H Huxley went one better: the cause of plentiful cats was old maids, and old maids were therefore the cause of the greatness of England!

Despite this promising example, that included both flora and fauna, most ecological work before the 1930s was concerned with the vegetable kingdom alone. Ecologists studied the distribution of plants and tried to establish the succession of flora that would occupy a particular habitat until a climax or equilibrium was reached: in southern England the climax (if nature was left to her own devices) might be a forest of beech, oak, or birch; in northern Russia, coniferous forest; in the American mid-west, grassland. The British Ecological Society was founded in 1913, having grown out of an earlier body, the Central Committee for the Survey and Study of British Vegetation established in 1904. By 1913 the National Trust had acquired Wicken Fen and Blakeney Point as nature reserves, both destined to become famous in the annals of ecology. For many years the great majority of the work published in the society's journal reported the impact of natural forces – climate, fire, altitude, slope, rabbits, and so on – on vegetation. The impact of man as polluter or farmer rarely came into consideration. There were few articles on smoke or the condition of rivers, and the word pollution did not occur in the index to the first 20 volumes covering the years 1913–32.

Animal ecology got little attention, which was hardly surprising, since it was much easier to study plants that remained in one place than animals that moved about. The first English book on animal ecology – by Charles Elton (several had already appeared in America) – was published in 1927 and reviewed in the *Journal of Ecology* by A G Tansley. Elton laid down some key principles of animal ecology, including the food-chain and the ecological niche. The food-chain was an essential tool for ecologists, who in the 1930s began to trace the sequence of events when chemical poisons like pesticides worked their way through a succession of creatures. The idea of the ecological niche provided a justification for attempts to preserve some species, for example in sites of special scientific interest. These are simple practical applications of concepts that have a much wider bearing in the eyes of ecologists. As the volume of research grew, the British Ecological Society established separate journals for animal ecology (1932) and for applied ecology (1964).

The development of ecological studies has altered the tone of biological thought. While the idea of evolution was dominant, biology had a dynamic forward-looking attitude; change, very slow change admittedly, was to be expected from the working out of evolutionary forces. Man himself could not escape from the pressure for change. As with other creatures, it was presumed, his origins went back through earlier ancestors to the primaeval slime. And according to the theory, the whole history of evolution would be demonstrable if the geological record were less imperfect. Ecology, on the other hand, paid less attention to time as an element in biology: it studied comparatively short-run problems – the struggle for existence within and between species in a forest or in a pond, for example – and tried to determine under what conditions equilibrium might be established or restored. It fostered a conservative rather than a dynamic outlook, and it was no coincidence that some ecologists spoke in accents remarkably like those of Edmund Burke. The influential American naturalist Aldo Leopold distrusted all meddling with the natural order:

> The outstanding scientific discovery of the twentieth century is not television or radio, but rather the complexity of the land organism. Only those who know the most about it can appreciate how little is known about it. The last word in ignorance is the man who says of an animal or a plant: 'What good is it?' If the land mechanism as a whole is good, then every part is good, whether we understand it or not. If the biota, in the course of aeons, has built something we like but do not understand, then who but a fool would discard seemingly useless parts? To keep every cog and wheel is the first precaution of intelligent tinkering. Have we learned this first principle of conservation: to preserve all the parts of the land mechanism? No, because even the scientist does not yet recognise all of them. . . . There is as yet no ethic dealing with man's relation to land and to the animals and plants which grow upon it. Land, like Odysseus's slave girls, is still property. The land-relation is still strictly economic, entailing privileges but not obligations. . . . A land ethic changes the role of Homo Sapiens from conqueror of the land-community to plain member and citizen of it. It implies respect for his fellow-members and also respect for the community as such. (Leopold, *Sand-county almanac*, pp 176–7, 218–20.)

Not all ecologists shared Leopold's aversion to change. Some like Elton recognised that any natural equilibrium was unstable, liable to be upset by fire or flood, or by climatic or geological change. Others argued that certainty was unattainable, and that it would be better, say, to tackle pollution on the basis of imperfect knowledge, rather than to take no action at all for fear of possible adverse effects on the natural environment. But these are dissident voices. Ecology has become for many people a faith that arouses passions too deep for the compromises of politics.

Since Copernicus, man's self-esteem has taken many hard knocks without being much dented. It was possible to brush aside the discovery that the earth revolved round the sun and not vice versa, for the earth seemingly remained the only living planet and man the master of it. When Sir Fred Hoyle wondered if other galaxies could field an eleven strong enough to play test cricket with the MCC, he was merely airing an eccentric hypothesis. Even Darwin's *Origin of Species* and *Descent of Man* proved to be less wounding than at one time seemed likely. Advanced churchmen found themselves able to fit the theory of evolution into their scheme of things, and convinced Darwinians like Sir Julian Huxley saw man as the central figure in the natural world; Huxley believed that it was only through man that the process of evolution could go on. Ethology and ecology appear to have succeeded where astronomy and evolutionary theory failed. Ethologists have made the shameful discovery that man is a less merciful creature than the wolf or the greylag goose. Ecologists, with geologists and economists, have shown the limits of man's power and the truth of the adage that, browbeat Nature as you may, she will have the last word. In ecology man is a powerful and disturbing influence but not necessarily the most important figure, or the most highly esteemed.[6]

THE ENVIRONMENTAL MOVEMENT

Scientific findings do not necessarily arouse much public concern and the level of interest in environmental questions has waxed and waned over the years, without ever coming to the top of the political agenda. The rapid growth of towns in early Victorian Britain provoked some sanitary reforms, but atmospheric pollution and the state of the rivers received comparatively little attention. Aesthetes from Wordsworth to William Morris and beyond protested at ugliness, noise, or the insensitive restoration of churches without much effect. Academic economists continued to accept the Malthusian analysis. But in the eyes of most people the Malthusian system had been discredited before the end of the nineteenth century, so far as the western world was concerned, by American wheat and family planning. The depletion of natural resources aroused some anxiety in America early in this century, but the British response to similar anxieties was much more tepid: the Liberal government instituted the Development Commission in 1909, but it remained a minor part of the machinery of state. After both world wars there appeared for a time to be a shortage of raw materials, but the anxiety was short-lived. After 1918 supplies recovered to such good effect that in the economic crisis of

1929–33 it was the plight of primary producers unable to sell their wares that aroused most concern. After the Second World War similar fears about raw-material shortages were similarly answered: by the late 1950s cheap oil from the Middle East was alleviating any shortage of coal, and the prices of other raw materials also fell as supplies increased. The third quarter of the twentieth century bore an uncanny resemblance to the corresponding period a hundred years before: rapid economic growth culminating in a crisis in the early seventies.

But if history repeats itself it is with variations. The environmental anxieties of the 1860s did not grow into a substantial movement of opinion, whereas the strong environmental anxiety evident by the end of the 1960s has been sustained ever since. That concern came about rather suddenly, but it certainly did not trouble the early years of the Labour government of Harold Wilson. Those years were spent formulating a national plan for economic growth, and the planners foresaw no difficulties in raw-material supply, and did not even mention the topic of pollution. The government created a new, grandly named department, the Ministry of Land and Natural Resources, which scarcely lived up to its title. In practice it confined itself to securing land for public housing, and when it was wound up in 1967 the minister, F T Willey, became second in command at the Ministry of Housing and Local Government with some loss of status but no real loss of scope or power. If the public had felt deeply anxious about the environment it is unlikely that a skilful politician like Harold Wilson would have abandoned so promising a title. In a few more years a ministry so named might have struck a chord in the public mind, but in the mid-sixties, as ten or twenty years earlier, questions of pollution and conservation were not newsworthy. The index to *The Times* newspaper had no entry for these topics in the last quarter of 1946, 1956, or 1966 and it was only towards the end of 1969 that *The Times* began to give more than a minimum of space to environmental questions. Coverage became much fuller in 1970 and the environment has continued to be newsworthy ever since – of course, in a frivolous world, not on the scale of fashion, politics, or sport.

There were other early signs of a growing concern with the environment. The Conservation Society, a quietly educational body now absorbed into the Environment Council, began its work in 1966. A new journal *The Ecologist* published in 1972 the *Blueprint for Survival* cautiously endorsed by such distinguished biologists as Sir

Julian Huxley, P Medawar and G Wynne-Edwards. It proposed to restrain economic growth and new technology, and somewhat inconsistently argued for a lessening of the powers of the state.[7] Despite its shortcomings the *Blueprint for Survival* like the Club of Rome's *Limits to Growth*, attracted a good deal of attention, though its proposals were far removed from the realm of practical politics. The Ecology Party founded in 1975 made little headway, either under that name or under the catchier title of the Green Party. Its one brief success was to take 14 per cent of the votes (and therefore none of the seats) in the 1990 elections to the European Parliament at a time when the Liberal Democrats were in some disarray. Subsequently support for the Greens has dwindled because the Liberal Democrats, the third force in British politics, have recovered their poise. Lack of electoral success, especially under the haphazard system that passes for parliamentary democracy in Britain, does not deny all influence to minority parties. It is often said, with little truth, that the infant Labour Party leavened the Liberal lump in the years before the First World War when Liberal governments were in office (1906–14). It would be possible to make a better case for the Green Party in modern Britain. The three main parties profess their devotion to the environmental cause, though two of them (it may be suspected) care more for economic growth. But when a prime minister whose leanings are towards the economics of Richard Cobden, speaks of our duty to posterity and of the dangers of the greenhouse effect, it is pretty clear that environmental questions have become serious politics.

This short introductory chapter has sketched some of the leading ideas that have shaped the environmental movement. The next four chapters consider problems of pollution – in air and water, and on land. The sixth chapter completes the first half of the book with some study of the efforts to preserve natural beauty, ancient monuments, and fine buildings. The second half of the book concerns itself with matters more strictly economic, with the inroads made by industrial Britain into exhaustible natural resources – coal, iron, base metals, and other minerals. It then examines ways to economise by recycling and the use of byproducts, and by variations in the quality of goods. Finally there is a short discussion of the meaning of economic growth in a world of finite resources.

REFERENCES

1. B Russell, *Authority and the individual* (Reith Lectures, 1949), p 93.
2. T R Malthus, *An essay on the principle of population* (1798 repr in facsimile 1966), p 90; 6th edn 1826, repr as 7th edn 1872, p 370.
3. For some account of Malthus's precursors, see Patricia James, *Population Malthus* (1979), pp 56–60.
4. R H Parker, *Management accounting: an historical perspective* (1969), pp 34–8; R de Roover, *Money, banking and credit in medieval Bruges* (Cambridge, Mass, 1948), pp 125 and 143n; A de Moivre, *Annuities upon lives* (1725), 4th edn 1752, pp 50, 82–3; de Moivre's *Doctrine of chances,* (ed posthumous 1756), which incorporates the treatise on annuities, was reprinted in facsimile in 1967.
5. John Richards of Exon, *The gentleman's steward* (1730), pp 8, 10, 86, 88; E Laurence, *A dissertation on estates upon lives and years* (1730), pp 12, 19, 30–40; select committee [s.c.] on church leases (Parliamentary Papers [PP] 1837–8 IX), qq 148–9, 157, 224–8; departmental committee on forestry (Board of Agriculture), *Evidence* (PP 1903 XVII); royal commission [r.c.] on coast erosion, *Second report* (forestry) (PP 1909 XIV); M Gane, ed, 'Martin Faustmann and the evolution of discounted cash flow', Commonwealth Forestry Institute, *Institute Paper* no 42 (Oxford, 1968) – this work includes translations of two of Faustmann's papers; Lords committee on the coal trade (PP 1830 VIII), pp 37–8; Haydn Jones, *Accounting, costing and cost estimation: Welsh industry 1700–1830* (Cardiff, 1985), pp 155, 169–71, 181; W Hoffmann, *British industry 1700–1950* (trans W O Henderson and W H Chaloner, Oxford, 1955), esp. pp 117–18.
6. E Haeckel, *The history of creation* (1873); *Evolution of man* (1879), I, pp 114 and 279; C Darwin, *Origin of species* (1859) repr 1968 in Penguin Classics, p 125; E J Kormondy, *Concepts of ecology* (Englewood Cliffs, NJ, 1969), viii; *Index to the Journal of Ecology* vols I–XX (1933); C S Elton, *Animal ecology* (1927, repr 1953), pp 20, 55ff; R A Leopold, *A sand-county almanac* (NY, 1949, 1966 edn), pp 176–7, 218–20; T Last, 'Effects of atmospheric pollution on forests and natural plant assemblages' (National Society for Clean Air [NSCA], *44th annual conference 1977*), p 16.
7. The National Plan, Cmnd 2764 of 1965; *Parliamentary debates,* 26 November 1964, col 214 ff; *Index to The Times* various dates [v.d.]; *The Ecologist,* vol 2, no 1, January 1972, pp 1–43; Margaret Thatcher, *The revival of Britain: speeches on home and European affairs 1975–88* (1989), p 274.

PART ONE

Pollution and Amenity

A winter's morning in the West End of London, 1924. Thick fog accompanied by severe pollution threatened the health and comfort of townsfolk at irregular intervals in the winter months, from the seventeenth century to the 1960s.

CHAPTER TWO

The Age of Smoke and Smells

Every economic process leaves behind a waste product. A coal fire, whether in a poor man's cottage or an industrial boiler, leaves behind ash, soot, and carbon particles, which either escape into the air if they are finely divided, or remain as solids if they are not. In addition, waste gases – oxides of carbon and sulphur mostly – will be discharged into the air unless means exist to detain them. The smelting of metals leaves a large residue of dross and slag and much smoke and fume. A modern power station or cement works will shower the neighbourhood with dust unless carefully managed. Oil refineries and chemical works threaten to give off poisonous and evil-smelling gases. A coffee roaster can fill a shopping street with smoke and smells. Even a performance of Beethoven's Eroica Symphony will generate a little dust as the horsehair bows of the string players gradually wear out. Pure air is therefore like liberty – only to be had at the price of eternal vigilance. And in practice some degree of impurity or pollution can hardly be avoided. In a state of nature the air will not consist simply of oxygen, nitrogen, carbon dioxide and the rare inert gases. It will contain dust and particles blown by the wind as well as a varying amount of water vapour. Atmospheric pollution occurs when man, volcanoes, or fierce winds substantially change the normal composition of natural air. This statement does not amount to a clear-cut definition, but in practice gross pollution, like the dog, can usually be recognised even if it is hard to define.

MEASUREMENT OF POLLUTION

Atmospheric pollution in Britain was probably at its worst towards the end of the nineteenth century. In 1881–5, the earliest years for which records exist, central London in the two winter months of December and January saw little more than one-sixth of the bright sunshine enjoyed in four small country towns.* At no time in the next forty years did London have as much as half the winter sunshine of these country towns, though the skies over the capital were getting slowly but steadily clearer. Even in the summer, central London suffered some loss of sunshine: in 1881–5 it lost a sixth of the sun it might have enjoyed in a state of nature. By 1920 the summer loss was imperceptible – a mere four per cent. By another test – the weight of solids deposited per square mile, per acre, or per square yard – London and other great cities suffered just as badly. The standard chosen to measure pure or nearly pure air was the deposit of solids in the pleasant spa town of Malvern. There the average monthly deposit per square mile amounted to five tons, of which less than a third was soot and sulphates from the burning of coal. Over half of the deposit appears to have been wind-blown earth and leaves, an irreducible minimum of impurities avoidable only on the moon. By the standards of Malvern the inhabitants of the great towns of England breathed air heavy with pollutants. Attercliffe, an industrial suburb of Sheffield, suffered most, with a deposit of 55 tons of solids per square mile. London and Manchester, lacking in heavy industry, had deposits of 38 and 32 tons. It is not easy to grasp the meaning of 55 tons of solids falling per square mile per month. To the inhabitants of Attercliffe it meant eight ounces of

Table 2.1 Mean monthly deposits 1914–16[1]

	(Tons per square mile)				
	Soot	*SO_3*	*C1*	*NH_3*	*Total Solids*
Attercliffe, Sheffield	9.6	7.2	4.0	0.20	55
London (8 stations)	5.9	5.5	1.7	0.30	38
Manchester	4.3	4.8	1.3	0.13	32
Malvern	0.4	1.1	0.5	0.04	5

Source: N Shaw and J S Owens, *The smoke problem of great cities* (1925), pp 64–6, 91.

* Oxford, Cambridge, Marlborough and Geldeston. Two of these places lie in East Anglia, where there is more natural fog than in London.

dust falling every year on every square yard of the district. Though much less dust than this would have invaded homes, the burden of cleaning imposed on housewives must have been severe.

When the measurements in Table 2.1 were made London was enjoying much more winter sunshine than it had thirty years before. It seems likely therefore that if the deposit had been measured at earlier dates, still higher figures would have been recorded. Industrial districts throughout Britain had their own local and special sources of pollution; and all shared in the coal smoke generated by private households. Wherever there were large numbers of houses there would be smoke from open fires and kitchen ranges. An isolated farmhouse or village would not generate enough smoke to annoy itself or others, for smoke, like sewage, became a serious problem only on a large scale. Coal production and consumption were at a comparatively low level until the end of the eighteenth century. Only in London was enough coal burnt before then to generate much smoke – and complaint. It has been estimated that British coal production amounted to about ten million tons in 1800. Thereafter, output grew decade by decade until it reached a peak of 287 million tons in 1913. An irregular decline followed, to less than 230 million tons by the outbreak of the Second World War. Output fell further during the war, but had recovered by the 1950s to the depressed levels ruling in the later 1930s. The output of coal is not a particularly good guide to the output of smoke. A better guide is the quantity of coal retained for home consumption after exports have been deducted. Exports grew from five per cent of output at the beginning of the nineteenth century to ten per cent in the 1860s and to 25 per cent in 1913. If coal for foreign bunkers is included in exports, the proportion rises to 33 per cent in 1913. The collapse of demand between the wars was largely due to lower exports and to the increasing use of oil-burning ships. In prosperous years, at least, home demand for coal held up well. With the virtual disappearance of coal exports during and after the Second World War, practically the whole output was available for home consumption. Full employment and a rapid rise of manufacturing output actually led to an increase in the home demand for coal that lasted for some ten years after the Second World War. The course of home consumption is therefore as shown in Table 2.2:

Table 2.2 Consumption of Coal in Britain 1800–1950[2]

	(*million tons*)
1800	10
1856	60
1869	97
1900	167
1913	189
1929	176
1938	181
1950	194

It does not follow that the amount of pollution rose to a peak early this century and remained at the same high level until about 1950. Much depended on the efficiency with which coal was burnt. It will be shown later that in manufacturing industry and the generation of power, considerable strides were made in the avoidance of smoke. First, it will be useful to look at domestic heating and cooking. Somewhat shaky estimates made by royal commissions suggest that in 1869 and 1903 just under a fifth of the coal available for home consumption was burnt in domestic grates and stoves. In 1940, before supplies were rationed, over a quarter (26 per cent) of coal was for domestic consumption.* By 1945 the proportion had fallen back to a fifth and it remained near that level for some years. In other words, the domestic consumption of coal, which amounted to less than 20 million tons in 1869, rose to 32 million in 1903 and to over 50 million in 1940, perhaps the peak year. It fell below 40 million tons again in 1944.[3]

THE DOMESTIC FIRE

At first glance more coal burnt seems to imply more pollution unless the efficiency of combustion improves. Of improved efficiency there was little sign. The Englishman was notoriously attached to his wasteful open fire although it belched out smoke into the open air. The historian Edward Gibbon expressed the popular view when he recalled his early days in Lausanne:

* Figures for 1940 and after include coal burnt in shops and offices whose annual consumption did not exceed 100 tons.

> I had now exchanged my elegant apartment in Magdalen College . . . for a small chamber ill-contrived and ill-furnished, which on the approach of winter instead of a companionable fire must be warmed by the dull invisible heat of a stove.[4]

Experts knew that the continental stove warmed a room more effectively for a given quantity of fuel. Even if the supply of air was no better regulated than in an open fire, there would be less smoke for a given quantity of useful heat. Estimates of the heat loss from the open fire varied from seven-eighths to nine-tenths, and it was no wonder that a German chemist scornfully dismissed the English open fire as a system of chimney heating.[5] It is unlikely that many Victorian householders had any inkling of the low standard of efficiency of their open fires, and they certainly had no intention of giving them up for continental stoves. In 1857 the General Board of Health appointed commissioners to report on the heating and ventilation of dwellings. The commissioners deliberately ignored closed stoves (while admitting that they were cheap, clean and efficient) because they tended to make a room stuffy. Englishmen, including the experts, preferred fresh air to warmth and were willing to take their chance over dirt and expense. Though sharing English prejudices, the commissioners were anxious to avoid smoke and encourage greater heat output per unit of fuel. They reported favourably on some open-fire grates operating by gravity feed and having a constricted throat to prevent a wasteful rush of air up the chimney. None of these grates came close to eliminating smoke, and a generation later a further body of investigators admitted that a satisfactory open grate did not exist, and that without one it would be pointless to attempt to control the output of domestic smoke.[6] By 1903 the Coal Smoke Abatement Society could envisage an ultimate ban on smoky open grates. It recognised, however, that central heating – the best antidote to smoke – was unpopular and rarely to be found even in offices: 'A government clerk finds he is terribly injured if he has not a fire in the grate to poke'.

Indifference to the cause of smoke abatement was apparent in all ranks of society, even among those who could well afford the latest and best equipment. The Duke of Westminster was an early and consistent advocate of smoke abatement but the matter did not trouble the Duke of Wellington in Apsley House. Dr H A des Voeux, a West-End physician and chairman of the Coal Smoke Abatement Society, did not mince matters: 'Upon my word the Duke of Wellington's house is like a factory chimney in the morning, in fact some of the factories do not emit so much smoke as the Duke of Wellington's house'.[7]

The design of domestic grates did not change substantially until concern about the evils of smoke became widespread in the 1950s. In 1948, 98 per cent of living rooms still had an open fire, or a fire combined with a back-boiler or a stove. Coal-fired cooking stoves were still being installed in many new houses between the wars, and a quarter of homes were still cooking by coal in the early 1950s.[8] At first sight it would seem that the war against smoke on the home front was being lost as the twentieth century wore on.

There can indeed be no doubt that the amount of smoke discharged from domestic chimneys was increasing until the Second World War. But by chance rather than forethought this did not lead to a corresponding growth of pollution and fog. The reason lies in the flight to the suburbs. It was already noticeable in the 1880s that Kew Gardens, then on the western fringes of the metropolis and about eight miles from Charing Cross, enjoyed almost as much winter sunshine as country places. The same observation applied to smoky Manchester in the early 1950s. Platt's Fields, about two miles out from central Manchester, tended to set the boundary of the area of thickest fog, and even on a clear day in the outer suburbs south of Manchester, a murkier air would be seen and breathed as the commuter left Platt's Fields behind him and came to crowded Rusholme.[9] Density of population naturally diminished in the suburbs, and with it there came fewer chimneys and less smoke to the acre. The spreading suburbs not only diluted their own smoke in a greater volume of fresh air, they also relieved the crowding of the inner suburbs. The London borough of Tower Hamlets, for example, now has a population of about 150,000. In 1900 four times as many people lived in the boroughs of Stepney, Bethnal Green and Poplar, which were merged to form Tower Hamlets in 1964. Inner London – the London of the Metropolitan Board of Works, the London County Council and the ILEA – housed 4.5 million people in 1900; the number had shrunk to 3.3 million by 1951 and is now down to 2.3 million. A similar pattern of population movement can be traced in all major British cities and in other industrial countries too. Dispersal of smoke necessarily follows dispersal of people and a little pollution widely spread replaces a great deal concentrated in a small area. Prevention would have been better than dilution, but there was little hope of preventing domestic smoke in the first half of this century.

INDUSTRIAL SMOKE

Industrial smoke was another matter. Early industrialists were careless in the use of coal and it is likely that most of the smoke and fogs that hung over early and mid-Victorian Manchester, Sheffield or London, came from boiler-chimneys, furnaces, gas works, railway engines and so on. Nowadays the motive power for industry takes the form of electricity, a clean fuel at the point of consumption. It is hard for us, accustomed to the centralised generation of power, to visualise the very different conditions under which a myriad of steam engines delivered power to nineteenth-century factories and workshops. It is true that water-driven cotton mills and corn mills went on being built and used with considerable success until well after 1850. But they were a small and diminishing proportion of industry. The future lay with the steam engine. A cotton or woollen mill, even a large one, rarely needed more than a few steam engines. It might employ some thousands of men, women and children, but it was a compact industrial enterprise and the belting driven from one boiler house could provide power for the whole plant. It was far different at a major ironworks. The Dowlais Ironworks at Merthyr Tydfil had 23 steam engines in 1837 and 63 in 1856. When the proprietor, Sir John Guest, gave evidence to a parliamentary enquiry into smoke prevention he argued forcefully that efficiency and smoke prevention did not go together. The demand for power in an ironworks was intermittent; when full power was needed, for example to roll iron rails, the extra demand could be met only by furious stoking – and clouds of smoke. A woollen mill, Sir John argued, faced no such intermittent demands for power; regular, comparatively smoke-free stoking would be technically feasible there. In the more hectic environment of an ironworks with its multiplicity of engines, a smokeless atmosphere was a luxury that he could not afford. In any case, he added, most of the smoke at the Dowlais works came from the furnaces, and the proposed bill exempted furnaces from control. Dowlais was not exceptional in this respect; it may not even have been the largest ironworks of the day. Sir Joseph Bailey, another ironmaster MP who gave evidence in 1845, was described by the chairman of the enquiry as the 'first leading ironmaster in Great Britain'. His works in Brecon and Monmouth consumed more than a quarter of a million tons of coal in a year. Less than a fifth of this huge quantity was needed for his powerful steam engines and like Sir John Guest he argued that iron and smoke were inevitably in joint supply. A generation later the Bowling ironworks of Bradford presented a similar picture. An integrated concern with its own coal and iron mines and blast furnaces, the Bowling works was famous for the high quality of its bar iron and its steel. The bar iron was worked up into plates or rails and the steel was largely destined for use in the Sheffield cutlery trade. The mines employed 2,000 men in 42 pits powered by 61 steam engines. The ironworks employed a further 1,000 men and 17 large and numerous small steam engines.[10]

Few industries were as lavish in their demand for power as iron-making, but every concern above the level of a small workshop was likely to need at least one steam engine. There are few figures for the number of stationary steam engines and they are all mere estimates. It is said that the textile industry in 1838 employed 3,000 engines. Witnesses called before the Coal Commission in 1870 put the total number of engines at between 50,000 and 100,000. Another authority in 1872 was even more non-committal, offering a range from 50,000 to 200,000. The steam power available in Britain probably increased fivefold in the next thirty years but not necessarily with any great increase in the number of engines, since larger and more powerful machinery was all the time coming into use. After 1900 electric power spread slowly through industry with a consequent reduction in the number of engines, but as late as 1952 Sir Hugh Beaver estimated that there were still 40,000 or 50,000 hand-fired boilers at work, all giving out more or less smoke.[11]

The Beaver Committee distinguished between hand-fired boilers and mechanical stokers because hand-firing was and always had been an erratic process. Whenever industrial smoke was in question, the pay, status and skill of the stoker came under discussion. The ideal stoker, it was agreed, would be a skilled workman, well-paid and not burdened with other duties. He would shovel at least 50 tons of coal a week, a little at a time. He would not smother the fire, but leave some of it red and therefore hot enough to burn up volatile gases and particles of carbon – the main constituents of smoke. In many firms, however, the stoker was ill-paid and had to oil the machinery and inspect the belting. His frequent absence from attending to the fire caused him to smother it with a thick layer of coal, resulting in clouds of black smoke. If he admitted a great deal of air in order to keep down the smoke, the boiler would cool, for the stoker would be guilty of the typical vice of the domestic fire-tender – chimney-heating. Some employers did pay their stokers well: Gotts, the Leeds blanket makers, were paying 24s. a week in 1845, a good wage for the time. But too often stoking was looked upon as rough, unskilled, manual labour and paid accordingly. The Manchester Steam Users' Association, a body concerned with fuel efficiency as well as boiler safety, employed a peripatetic inspector to train its subscribers' firemen from the 1860s onwards. In the Second World War and after, the need to conserve scarce fuel led to a larger effort to train stokers, and over a ten-year period, 46,000 attended classes in fuel efficiency. Attendance at classes and willingness to learn have never been synonymous. One hard-bitten engineer advised employers to avoid experienced stokers with 15 years' service in the Royal Navy. They would prove unteachable, whereas a ploughman, untroubled by preconceived ideas, would do as he was told.[12]

The total quantity of coal shovelled by stokers, probably with slowly increasing efficiency, must have gone on rising throughout the nineteenth century. Mechanical stoking, like mechanical coal-cutting, has a long history, but it made relatively little impact in Britain until the late nineteenth century. An early device patented by Charles Wye Williams attracted the praise of Andrew Ure, the scientific populariser, but proved disappointing in practice – a common enough failing of new ideas in the early stages of development. Other devices patented by Robertson, Jukes and Crampton fared little better. Some manufacturers took them up with enthusiasm for a few years and then reverted to the old ways. The main obstacle seems to have been that the mechanical stokers were too flimsily built and did not stand up to the arduous operating conditions without heavy bills for repairs. The inventors and their supporters found it hard to understand why the obvious benefits of smoke control did not attract more attention from industrialists, and they tended to blame the conservatism of millowners when, given the unreliability of the mechanical stoker, caution was a rational stance. It was not until the 1890s that sturdier construction reduced the costs of maintenance and made mechanical stoking a practical proposition. By 1907 Babcock and Wilcox, the leading firm of boilermakers, was fitting chain-grate mechanical stokers to all its large boilers. The days of manual stoking were now numbered but it naturally took many years to replace old boilers with new, or to turn over to oil, or to electric power.[13]

Boilers were not the only industrial source of smoke, though they were the most widespread. It is the fashion to romanticise steam railway engines, but any traveller old enough to remember the smoke, the smuts and the sulphurous fumes of the railway engine, will temper his enthusiasm for the good old days of steam. The railway engine had a character of its own no doubt, but it scattered a stiff dose of pollution the length and breadth of the land, much of it given off in towns where other sources of smoke were most numerous. Among more localised sources of industrial smoke were blast furnaces, coke ovens and pottery kilns. As the Welsh ironmasters pointed out in 1845, boilers accounted for only a small part of the smoke given off at a major ironworks. Blast furnaces – like coke ovens – were rarely capped in mid-nineteenth-century England. Their voracious appetite for coal or coke led to a corresponding output of smoke, accompanied, where combustion was imperfect, by large quantities of carbon monoxide. Gradually ironmasters overcame their fear of closed blast furnaces and by the end of the century hardly any open ones remained. But the making of metallurgical coke was still being conducted in a cloud of smoke. The early eighteenth-century demand for coke was met by heaping up coal in the open air and driving off the valuable byproducts – tar, ammonia, carbon monoxide and other gases – amid thick black smoke. When ironmasters adopted the beehive oven the smoke nuisance was not much abated, for it was not uncommon for smoke and other waste to be discharged at a height of only ten or fifteen feet. Small wonder that the members of a royal commission visiting Tyneside in 1877 'observed a canopy of black smoke almost everywhere'. By the end of the century the beehive oven was only slowly giving way to continental-style retorts at collieries and ironworks, and the process took many years to complete. When the coal industry was nationalised in 1947 the National Coal Board acquired, among much other antiquated lumber, 26 coke ovens built in 1882. Although they were said to be in relatively good condition they were almost certainly of the beehive type that gave off black smoke in profusion. Sheffield and northeast England as major centres of coal-mining and iron and steel making were notorious for their smoky atmosphere.

The Potteries had a reputation every bit as bad, and the coal-fired bottle kiln did not come under serious challenge until after the Second World War. Although an electric furnace was demonstrated to an apparently sceptical audience in Stoke as early as 1906, it was 1927 before such a furnace was installed. The first tunnel-kiln fired by gas set to work in 1911, but the widespread use of gas did not begin for another twenty years. There were still about 2000 coal-fired kilns at work in 1939 and the inhabitants of the pottery towns continued to breathe a heavily polluted air. Indeed, the amount of smoke in the air, winter and summer, tended to increase until the output of pottery was restricted during the second year of the war. The skies over Stoke only began to clear after 1945 with the widespread adoption of firing by gas or electricity. By 1960, the winter concentration of smoke had fallen to about half of the prewar average. But the five towns remained a smoky place to live.[14]

ACID, DUST AND SMELLS

Smoke and winter fogs were the most obvious signs of atmospheric pollution. In some towns, and parts of towns, there were other pollutants as unpleasant and sometimes more dangerous to health. They can be summed up as being in one of three classes: acids, dust, and a miscellaneous collection of bad smells – effluvia was the ponderous and euphemistic word preferred by the Victorians. The acid emissions that arose from the burning of coal were much the largest. Carbon thoroughly burnt turns to carbon dioxide, and large and growing quantities of this gas were emitted in the nineteenth century. It aroused little concern at the time since carbon dioxide does not dissolve very readily in water to form carbonic acid, and the acid is in any case weak. A handful of scientists – Fourier in the 1820s and Arrhenius in 1896 – pointed to the dangers of the greenhouse effect, but this problem aroused little interest before the late 1980s. Sulphur is always present in coal, and usually accounts for between one and two per cent of total weight. Like carbon dioxide, sulphur dioxide and sulphur trioxide do not combine so freely with water as to deposit sulphurous and sulphuric acid in large doses close to where coal has been burnt. As recent experience of acid rain has shown, high chimneys successfully distribute sulphur compounds over a very wide area. Nevertheless the damage to masonry and ironwork from corrosive acids in the air was obvious in nineteenth-century towns and the acrid taste of winter fogs was an unpleasant experience.* Nobody in authority saw fit to protect the public from the acids arising from the combustion of coal. Acid emitted in particular industrial processes was another matter.

* In my recollection more unpleasant in Manchester than London, probably because of the high sulphur content of Lancashire coal.

The most famous example of industrial pollution by the emission of acid occurred in the alkali manufacture, the largest branch of the heavy chemical industry in nineteenth-century Britain. The principal product of the alkali manufacture was soda or sodium carbonate. Soda was valuable in glass-making, and as a cleaning agent in its own right; and caustic soda, a derivative, was an ingredient of soap. Put simply, the manufacture of alkali (soda) by the Leblanc process took place in two stages: first, salt combined with sulphuric acid to produce sodium sulphate and an unwanted byproduct – hydrochloric acid. Sodium sulphate was later burnt with chalk and coal to produce soda and another troublesome byproduct – calcium sulphide. Of the two byproducts, hydrochloric acid was by far the more pernicious. It was given off from open furnaces as a gas and since it readily dissolved in water, it was little diluted before it came to earth. The atmosphere in towns like Runcorn was distinctly sharp as recently as the early 1950s when alkali manufacture was much reduced. In the nineteenth century Runcorn, Widnes and St Helens, where the alkali manufacture had settled, were notorious for the impurity of their air. For several miles on the leeward side of the alkali works trees died and crops and grass grew poorly. There were other important centres of the industry in Glasgow and on Tyneside, and some smaller works in London, Bristol and elsewhere.

In 1836, William Gossage, whose works then lay at Stoke Prior in Worcestershire, patented a device for condensing the hydrochloric acid in towers packed with brushwood (later coke was used) down which water trickled, absorbing the gas and producing a weakish solution of acid. If most manufacturers had adopted this process, and if they had then found a use for the hydrochloric acid, the atmosphere on Merseyside and elsewhere would have been sweeter. Unhappily Gossage's towers were not much seen in the industrial landscape and when they were installed the acid was often discharged into the nearest river. Not only did the acid further pollute some already unwholesome waters, it also contributed to another form of atmospheric pollution. This arose from the combination of hydrochloric acid (either raining down from the skies or discharged into rivers) with the calcium sulphide that also resulted from alkali manufacture. The combination of these two chemicals released sulphuretted hydrogen. The smell (like rotten eggs) sometimes spread far and wide over south-west Lancashire. The nuisance became so bad that in 1876 a royal commission was appointed to report on this and other noxious vapours. A Liverpool solicitor P F Garnett living at Aigburth, seven miles from Widnes, kept a record of stenches arising from alkali waste. He noticed the stench 24 times in 1875 and 29 times in 1876. On one day in April of that year it was perceptible in Castle Street, Liverpool, some twelve miles from its point of origin.[15] Until a use could be found for the hydrochloric acid, the solid alkali waste, or both, there were only poor prospects of ending this particularly unpleasant kind of atmospheric pollution.

Though the most extensive and the best known source of pollution by hydrochloric acid, alkali manufacture did not stand alone. Wherever salt or hydrochloric acid was used in industry, the danger of unpleasant emissions arose. The rich salt deposits of Cheshire have long been a mainstay of the British chemical industry, and many salt-pans were built to evaporate the brine pumped up from beneath the Cheshire plain. Not all the salt had to be pumped, for there were also enormous deposits of rock salt to be mined in the usual way. But brine was important and in the heyday of the saltpans more than 1,000 were at work. Unless carefully controlled, the process gave off substantial quantities of hydrochloric acid gas, admittedly in the neighbourhood of relatively small Cheshire towns like Northwich and Middlewich where fewer people suffered annoyance than from the alkali manufacture on Merseyside.[16] Another source of emissions was the Yorkshire shoddy trade. The shoddy trade reprocessed woollen rags by tearing them to shreds and spinning the product into an inferior yarn. This yarn, either alone or combined with yarn made from new wool, was then woven into cloth, not all of it deserving the name of shoddy. Problems arose when the rags were mixture fabrics containing cotton or linen as well as wool. The vegetable fibres stubbornly refused to submit to reworking and had to be eliminated. Luckily they would dissolve in a weak solution of sulphuric acid or in hydrochloric acid gas, leaving the wool unscathed. This process – known as carbonisation because it charred the unwanted fibres – was introduced in the early 1850s, not only in the shoddy districts of the West Riding, but also in London, the centre of the trade in rags. The product, known as extract, was disliked in the shoddy trade because its felting qualities were inferior to those of yarn made from pure woollen rags, and a large part of the output found its way to foreign mills. The manufacturers, until advised by government inspectors, made little effort to condense the hydrochloric acid gas and in 1884, shortly after inspection had begun, only one works had a condensing tower.[17]

A further source of annoyance from hydrochloric acid gas arose in the salt-glazing of pottery. Henry Doulton, a leading potter of the day, with works in London and St Helens, used salt glaze extensively and by rule of thumb. His works used much more salt than was necessary and therefore emitted large quantities of acid gas. In Lambeth he met doughtier opposition than in St Helens where in his opinion 'our works would not be considered a nuisance'. St Thomas's Hospital and Lambeth Palace had higher standards. No sooner was the Archbishop's silver cleaned than acid from the Doulton pottery tarnished it again. Archbishop Tait was so concerned about the nuisance and the danger to health that he gave oral evidence to the Royal Commission on Noxious Vapours. St Thomas's Hospital was also concerned about the acrid smells generated by the pottery, but was even more troubled by the stench of rotting vegetation awaiting shipment from nearby wharves to farms in the neighbourhood of London. For Lambeth was a collecting centre for much of the rubbish of London, including the refuse of Covent Garden. With all due deference to the physicians of St Thomas's, it is doubtful if cabbage leaves were as dangerous to health as hydrochloric acid gas. Trees planted in 1871 on the Albert Embankment close to Doulton's works were mostly dead by 1878 whereas trees planted on the Victoria and Chelsea Embankments on the opposite side of the Thames made satisfactory progress.[18] It seems unlikely that the mere smell of bad cabbage, however disagreeable, would have killed a reasonably healthy tree, whereas there was ample evidence from Lancashire of the damage done by acid gases released from the salt used in alkali works.

Not much could be done about the sulphur dioxide given off when coal was burnt. Had the sulphur content of coal averaged five per cent instead of one or two per cent, matters might have been different. There were industrial processes that did involve high concentrations of sulphurous fumes and these did give rise to concern and, eventually, to remedial action. One such process was the roasting of pyrites. Iron pyrites or fool's gold (FeS_2) in its impure commercial state contained too little iron to interest nineteenth-century ironmasters. For chemical manufacturers, however, its sulphur content was valued as a cheap alternative to pure Sicilian sulphur. Roasting the pyrites oxidised the sulphur to sulphur dioxide, from which it was a short step to sulphuric acid. Manufacturing chemists could not afford to let too much of the sulphur escape into the air though some loss was unavoidable except with extreme care. It was far otherwise with the roasting of copper ores ($CuFeS_2$) in the Swansea Valley. There the smelter aimed to recover copper; if there was also a cheap and simple way to recover sulphur, well and good. In fact there was not, and large quantities of sulphur dioxide along with many other kinds of industrial waste combined to turn the Swansea Valley into one of the most polluted landscapes of the world. Vivians, one of the largest smelting firms in the valley, made gallant efforts at considerable expense to lessen the nuisance. They even tried the effect of long flues and high chimneys, but forty years later in 1862 neither they nor other smelters persisted in these attempts, which on a commercial view did not pay. By contrast in St Helen's there was a ready market close at hand for all the sulphuric acid that could be made. By simply roasting copper pyrites it was possible to recover the sulphur and be left with a valuable byproduct, impure copper ready for further refinement. In the wet-copper process, a branch of the alkali manufacture, calcined pyrites were first roasted with salt to produce alkali and copper chloride; the copper chloride was then dissolved in water and pieces of scrap iron were added. The chlorine transferred to the iron, and copper was precipitated. Any process that heated salt fiercely was liable to produce noxious fumes and the wet-copper process was no exception.[19]

The acid emissions recorded above are, as it were, specimen charges against industry. It would be tedious to draw out a full list of the offences against clean air and more interesting to consider a different kind of pollution – the nauseating smells that were a common enough feature of urban life in Victorian England and are still sometimes encountered. The acid pollutants so far enumerated were largely of inorganic origin. Those calling for consideration next were mainly organic. In the country, if all else fails, a compost heap and an earth closet at the bottom of the garden are adequate receptacles for domestic refuse and human excrement; wood ash provides a valuable source of potash for the garden, and the ashes from coal mend paths.

In a crowded town more elaborate arrangements are needed, as was only too plain to doctors, Poor Law reformers, and local authorities by the 1830s. The water closet is said to have been invented by Sir John Harington, godson of Queen Elizabeth I. Joseph Bramah (1748–1814), a well-known hydraulic engineer, has a better claim to this useful invention, but even he did not live nearly long enough to see his brainchild widely adopted. For the water closet had to wait for main drainage and a regular supply of piped water before it became truly serviceable. Where land is available as well as piped water, a cesspool system is adequate, but in towns the cesspool system and the water closet did not go well together. Constant piped water was rarely to be found in English towns before the middle of the nineteenth century and a system of main drainage was at least as long delayed. The trunk sewers of London, for example, were not completed until 1864, and a good many years passed before all the houses in the capital were connected to them, and still longer before water closets fully replaced older methods of sanitation. In Manchester the corporation actively discouraged the water-closet system on the advice of their first medical officer of health, John Leigh, who served from 1867 to 1888. Other towns welcomed the water closet but like Manchester and London could go no faster than main drainage and piped water systems allowed. In Leeds, for example, privies and cesspits outnumbered water closets by three to one in the 1860s. By 1914 the victory of the water closet was assured but still incomplete. The largest towns had made most progress: in Manchester only 4,000 pail closets remained out of over 100,000 at one time in existence. Birmingham too had only a few pail closets left. In smaller towns there was a wide variety of systems: the ashpit closet could still be found in Boston, Brierley Hill, Marple and Sunderland; the sanitary pail in Belper, Bilston, Mossley and Northwich; some privy middens lingered in Chesterfield and Skelmersdale, and Eye and Thurlstone (in the West Riding) confessed to having earth closets. A few rural authorities still made no provision at all for the collection of night-soil or other refuse because, as Glemsford in west Suffolk explained, most householders had gardens and no great difficulty would be experienced in getting rid of household waste.[20]

Night-soil or human excrement was the most offensive but not the only unpleasant waste product that troubled urban households. Bones, fat, peelings and cabbage leaves all had to be somehow disposed of, as well as ashes. The linc of lcast resistance was to pile up rubbish in a heap and wait for the contractor to remove it. A refinement on this practice was the ashpit, a fixed container at first made of wood. Into the ashpit went ashes, vegetable and animal refuse, in fact all household waste. At intervals contractors or municipal dustmen would shovel out and cart away the contents. Even if night-soil was not tipped into the ashpit, the nuisance must have been considerable. One sanitary inspector in Leeds put it at the top of his list of sanitary evils, when he appeared before the rivers pollution commission in 1866. The commissioners were anxious to get him to say that the state of the River Aire was abominable – which it was. But he thought it one of the lesser nuisances: 'I would put the ashpit first, the pigs second and the bad drainage third'. In the 1880s galvanised dustbins began to replace ashpits, and by the outbreak of the First World War, the dustbin like the water closet, had mostly ousted less effective systems. They lingered in old buildings in London, however, and were not forbidden in new buildings there until the 1930s. An otherwise respectable mansion block in Battersea, built in 1898, still had chutes and ashpits after the Second World War. One day an unlucky dustman found himself crawling with mice as he shovelled out the contents. Only after this incident did the borough council order the ashpits to be replaced by dustbins.*

Evil-smelling though they were, the contents of ashpits had some value. The contractors who emptied them accumulated huge heaps of refuse elsewhere, waiting to be sorted and barged or carted to nearby farms. Charles Dickens did not record that Mr Boffin's 'Mounds' stank, but there is no doubt that other contractors caused considerable offence. The concern of St Thomas's Hospital has already been mentioned, but a much greater enormity existed in Plymouth. A chemical manure manufacturer there felt that his works were being unfairly blamed for a stench that really came from a gigantic dungheap nearby. As Plymouth's medical officer wrote in 1874:

> On proceeding to inspect [the source of the stench], I found a vast heap of reeking filth which I never remember to have seen equalled for character and quality, either at home or abroad during my whole life. I would say at a low estimate there must have been 3,000 tons of town soil deposited on the site, and to aggravate the effects of such a vast aggregation of stinking matter, some thirty pigs were busily engaged in nuzzling up and devouring the titbits of offal contained therein. (R.c. on noxious vapours, 1878, qu 9048ff.)

Three years later the nuisance was as bad as ever. Some asserted that the smell of the heap carried no further than a quarter of a mile and that it was the chemical manure works whose odours could be detected two miles away. It is impossible to arbitrate between these conflicting views a hundred years later, but neither casts a flattering light on the sanitary condition of Plymouth in 1877.[21]

---

* Information from Harry Taylor, a long-standing resident in Avenue Mansions. Ashpits often caught fire when hot ashes were shot into them.

The pigs on the Plymouth dung heap remind us that Victorian England had by no means shaken off the medieval association between town life and agriculture. It was notorious among sanitary reformers that pig-keeping added to the squalor that surrounded the inhabitants of North Kensington in the 1840s. Cow-keeping was a common practice in large towns until railways and refrigeration made the long-distance transport of milk feasible. Town cows, stall-fed on brewer's grains and cattle cake, yielded rich milk, no doubt at some cost in smells to the neighbours. More widespread, but perhaps less offensive, was horse manure. Never had Britain reared and supported so many horses as in the nineteenth century. The railway that destroyed their long-distance hauls more than offset the loss by generating short-haul traffic on an unprecedented scale. Equally unprecedented was the scale of horse-droppings. In this as in much else London stood out, and around 1900 perhaps a million tons of horse manure were cleared from its streets and stables annually.[22]

Some of the most nauseating smells in Victorian England arose from the chemical treatment of organic matter. Bones, blood and excrement all had potential value as fertilisers. Animal fat offered lubricants and soap. Sugar refining sometimes used materials that gave off nasty smells. And the manufacture of cement called for heat to be applied to dredged material containing organisms – with unpleasant consequences for people living in the neighbourhood. At its simplest, and scarcely worthy of the name of chemical manufacture, was the business of drying night-soil for use as fertiliser. Few trades can have been more unpopular with local residents, but it was a trade quite widely practised in the second half of the nineteenth century when the water closet was far from displacing rival systems. In a slightly more elaborate process night-soil could be treated with sulphuric acid to produce the delicately named 'poudrette'. This manufacture was finally discontinued only on the eve of the Second World War, though the fumes given off had been denounced as dreadful seventy years earlier.

The treatment of bones in the manufacture of phosphates gave rise to more than one offensive smell. First the bones were boiled to extract fat and materials for glue-making: this in itself was a noisome business. Secondly, sulphuric acid dissolved the softened bones, often under primitive conditions that allowed the escape of acid fumes. One of the offending firms was the celebrated works of J B Lawes, founder of the Rothamsted agricultural experiments. His works at Barking Creek began operations in 1857 and were still subject to complaint by the sanitary authorities of Plumstead twenty years later. Tallow melters, boilers of blood and tripe, fell mongers and tanners, were all regularly denounced and condemned in evidence before official enquiries and enumerated in lists of offensive trades laid down in the Public Health Acts of 1875 and 1936. Unpleasant smells sometimes defy the best efforts of modern technology: in the hot dry summer of 1976 smells from a modern factory processing animal products on the banks of the Exeter Canal could be detected over a mile away in the city centre. And in ordinary weather it was not uncommon for its peculiar odours to penetrate into houses at a distance of a quarter or half a mile. There can be no doubt that this factory was a paragon of good industrial practice by comparison with many small and carelessly conducted works operating in the second half of the nineteenth century. London, for example, had over two hundred 'noxious premises' in 1878, not counting slaughterhouses, of which there were over a thousand. Among these offensive trades there were six blood boilers or driers, 15 bone boilers, 27 manure manufacturers and 55 tallow melters. It was not only the poor of Wandsworth, Battersea and Poplar who suffered from the presence of these works. In 1862 newly developed Camden New Town, home to barristers and other well-to-do people, was similarly afflicted. The offensive trades of Belle Isle, Islington, had been established there in open country before the new housing had come to Camden.[23] In the absence of planning rules there was nothing but good sense to stop houses being built and let close to noxious businesses. In North London both builders and lawyers apparently lacked the necessary foresight.

Protests against atmospheric pollution, whether by smoke or otherwise, were nearly as old as pollution itself. In 1392 the City of London ordered butchers to remove their offal and garbage to convenient places south of the Thames in Southwark. And by the seventeenth century London's reputation as a place of fogs and smoke was well established. The diarist John Evelyn is the best-known pamphleteer to protest against the smoke and smells of London. His *Fumifugium* published in 1661 denounced:

> that hellish and dismal cloud of sea coal which is not only perpetually imminent over her head . . . but is universally mixed with the otherwise wholesome and excellent air that her inhabitants breathe nothing

> but an impure and thick mist. . . . And the weary traveller of many miles distance sooner smells than sees the city to which he repairs.

He added a further complaint that may still strike a chord in London audiences:

> For is there under heaven such coughing and snuffling to be heard as in the London churches and assemblies of people, where the barking and spitting is uncessant and importunate?

His remedies for this state of affairs were few: banishment of noxious trades to the east, an end to the urban burial of the dead (only achieved in the 1850s) and the planting of sweet-smelling trees, shrubs and flowers. The ponderous *Fumifugium* was reprinted with a lively but anonymous introduction in 1772. In the meantime London had acquired a new batch of noxious trades to add to those listed by Evelyn. Brewers, soap-boilers, dyers and lime-burners had been joined by glasshouses, foundries, sugar bakers and waterworks.

REMEDIES FOR ATMOSPHERIC POLLUTION

Another fifty years elapsed before public discontent resulted in a first shot at the statutory control of smoke. The Smoke Prohibition Act of 1821 tried to encourage the prosecution of a public nuisance from smoke by allowing for the recovery of costs. The act did not extend to colliery engines, however, or to furnaces. It added little if anything to substantive law and simply served to state that Parliament was against smoke, rather than for it. Some local acts regulating smoke followed. None had much effect. The Derby Act, for example, imposed fines if a chimney emitted smoke for a week; as chimneys hardly ever smoked for more than half an hour at a time and fires were often put out at night, there was no problem complying with this particular law. The Leeds Improvement Act of 1842 gave birth to a famous formula – 'best practicable means'. Under the terms of this act it was a defence to show that the best practicable means had been used to prevent smoke. When the science of smoke abatement had scarcely been established, the formula provided ample scope for a successful defence against prosecution. But that was not all. Leeds followed Derby in providing penalties 'for and in respect of every week' during which the annoyance continued. Under such a law it would be a miracle if a prosecution succeeded.

During the 1840s, two select committees enquired into the smoke problem, and Lyon Playfair and Sir Henry de la Bèche, chemist and geologist respectively, were commissioned to write an expert and official report. Various private Members' bills failed to become law at that time, and it was not until 1853 that a London act was passed. It provided penalties where steam engines and other furnaces (except in potteries and glasshouses) created a smoke nuisance; 'best practicable means' provided sufficient defence. The act applied to steamboats plying above London Bridge, and after an amending act in 1856, to shipping below the bridge. Lord Palmerston as Home Secretary in 1853 piloted the bill through Parliament. It was not an arduous task: the only debate, a brief one, was in committee where a couple of members urged him to drop the bill on the grounds that manufacturers could not comply with its provisions. Palmerston, the only leading statesman of the day to take any interest in the smoke question, had little patience with them:

> Here were a few, perhaps a hundred gentlemen connected with the different furnaces in London, who wished to make two million of their fellow inhabitants swallow the smoke which they could not themselves consume and who thereby helped to deface all our architectural monuments, and to impose the greatest inconvenience and injury upon the lower class. Here were the prejudices and ignorance, of a small combination of men, set up against the material interest, the physical enjoyment, the health, and the comfort of upwards of two million of their fellow men. He would not believe that Parliament would back these smoke-producing monopolists. (*Parliamentary debates*, 9 August 1853.)

Enforcement of the law fell to the Metropolitan Police, who at least for a while, took their duties seriously. In 1854 they secured over 150 convictions, most for smoke offences, a few for offensive trades; about half the offenders came from Southwark, reflecting the zeal of that division of the Metropolitan Police rather than Southwark's importance in the industry of London. By 1861 there were said to be nearly 8,000 furnaces in London (one-fifth of them in Southwark) capable of consuming their own smoke. Nevertheless, the number of convictions under the smoke acts rose to nearly 700. In 1866 the Home Secretary, then Sir George Grey, circularised the large towns of England and Wales enquiring about enforcement of the local laws against smoke. Only Liverpool appeared to be enthusiastic for enforcement, having secured 142 convictions in 1865. Birmingham and Sheffield were also making an effort, but in Manchester, Leicester, Leeds and Newcastle-upon-Tyne little was achieved. The law was admitted to be a dead letter and 'without perceptible effect' in the potteries. Swansea did not bother to reply to the circular! In London the law continued to be enforced – there were more than a thousand convictions in the five years 1877–81, mostly for a first offence. The inconvenience to business cannot have been serious for the fines commonly ranged from five shillings to five pounds and more often than not fell below the minimum fixed by Parliament.[24] The Public Health Act, 1875, gave all local authorities the opportunity to combat the nuisance of industrial smoke if they were so inclined. But the blackened buildings, the fog and the weight of dust and soot that was still falling on English towns when measurement began early this century, suggest that neither local nor central acts did much good. It is probable that economic and technical changes did somewhat more to reduce atmospheric pollution in the period 1840–1950.

The force of the law came down rather more heavily on emissions of acid, dust and unpleasant smells. Smells, like smoke and the poor, had always been with us, it is true, but not on the same scale as in the neighbourhood of a Victorian manure works. Acid and dust emissions could certainly not be classed as nuisances so old that men would accept them without protest. Effective control of these sources of atmospheric pollution began with the Alkali Act of 1863. By then the alkali trade was consuming a quarter of a million tons of salt a year, quite enough to generate considerable discomfort for men and vegetation downwind of the alkali works. In May 1862 the Earl of Derby, a great Lancashire landowner whose seat, Knowsley Hall, lay between St Helens and Liverpool, moved in the Lords for a committee of enquiry into noxious vapours. The motion passed, the committee set to work, and its report was ready in three months. Within a year the government had framed and Parliament had accepted a bill to regulate not nuisances or noxious vapours in general, but the alkali trade alone. An amendment pressed by opponents of the bill (including John Bright)* allowed as a defence, the carelessness of workmen – contrary to the general rule that principals are responsible for the conduct of their agents. These restrictions

* John Bright was still carping at the Factory Acts in the 1880s.

disappointed Lord Derby who feared, mistakenly, that they destroyed the usefulness of the act. The Alkali Act had two main provisions: first, it insisted that alkali manufacturers condense at least 95 per cent of their hydrochloric acid gas; and second, it provided for the appointment of alkali inspectors. The alkali trade was by no means averse from regulation and inspection; though there were some manufacturers who resented any interference, many recognised the need for some control of a growing evil – as long as the inspectors were men of standing appointed by central government. If the trade had to submit to control, the pain would be less if manufacturers were dealing with salaried gentlemen, like the factory and mines inspectorate, rather than with men of the lowly stamp of sanitary inspectors, ill-paid employees of a local authority. Parliament met the trade's wishes in full. The inspectors were appointed by the Board of Trade from the ranks (then rather thin) of Britain's scientists, and the first chief inspector was a distinguished Scots consulting chemist R Angus Smith, who had his office in Manchester. Smith kept his consultancy, and the Board of Trade noted with satisfaction that it got quite as much work out of him as out of a full-time inspector, and that no pension was payable, Smith being technically a part-timer. A gentle scientist – his friends thought he should have been called Agnus – Smith served as chief inspector for twenty years, and men trained under him held the chief inspectorship until 1920. His methods therefore moulded policy for the first fifty years of the Alkali Acts, and have strongly influenced it ever since. As he explained in 1877: 'It is better to allow some escape occasionally than to bring in a system of suspicion, and to disturb the whole trade by a constant and irritating inspection.' (R.c. on noxious vapours, 1878, qu 12256.) In his view a manufacturer who followed the advice of the inspectorate and installed equipment costing (say) £2,000 was effectively submitting to a very severe fine. To persuade rather than to bully was his chosen policy, and manufacturers rarely found themselves in court. These equable methods came in for harsh criticism from the royal commission on noxious vapours (1878) but the inspectorate went its own way, unperturbed. Manufacturers did more or less comply with the promptings of inspectors, often at considerable expense. The early attempts at measuring the escape of gas after condensing towers had been installed seemed to show that manufacturers could easily keep within the permitted limit of five per cent. Later measurements, however, suggest that these early results were unduly optimistic. In any case the amount of salt decomposed in the alkali trade doubled between 1862 and 1878 so that emissions would have increased even with full enforcement of the law. Moreover the inspectors had no control over the disposal of the condensed hydrochloric acid and it was this nuisance in particular that had led to the appointment of a royal commission.[25] The best that can be said for the early Alkali Acts is that without them the pollution would probably have been a lot worse. Yet it is doubtful if a harsher stance would have achieved more, given the rapid expansion of the trade, the power of the industrial interests concerned, and the general unwillingness to check economic growth.*

It was an anomaly that the 1863 act regulated the alkali trade and left other trades free to do as they pleased. A further act in 1874 did nothing directly to end the anomaly, though by noticing other noxious gases – defined as sulphuric and sulphurous acid, oxides of nitrogen, chlorine, and sulphuretted hydrogen – it brought closer the regulation of other trades in due course. The act reiterated the standard set in 1863 – condensation of at least 95 per cent of hydrochloric acid gas – and added a further restriction, that no more than one fifth of a grain of such gas should be emitted in each cubic foot of air or smoke.** At the same time the famous if elastic formula 'best practicable means' became the test applied for regulation of other noxious gases. Interpretation of the formula was left to the expert knowledge, judgement and tact of the inspectorate. Much criticised over the years, 'best practicable means' is in principle neither a better nor a worse guide than numerical standards. Under a careless administration it would allow gross pollution; given effective technology rigorously applied, it could result in clean air. Under the alkali inspectorate it was a source of gentle but steady pressure for a slow improvement. Modern industrialists seem to prefer best practicable means. It is therefore a reasonable inference that fixed standards would in practice provide better safeguards for the environment. The only new process regulated by the 1874 Act was the 'wet copper process' of alkali manufacture. There had hitherto been legal doubts as to whether this process fell within the limits of the Alkali Act.

A much greater change occurred in 1881 when the Alkali etc.

---

* Contemporaries would have used the roughly equivalent term 'economic progress'. Whether such terms are truly appropriate is a contentious matter.

** One fifth of a grain per cubic foot = 0.475 mg per litre; 7,000 grains = 1 lb.

Works Regulation Act became law. The 'etc.' indicated that some other chemical works were now to be controlled, as recommended by the royal commission on noxious vapours. Works manufacturing sulphuric or nitric acid, chemical manure works, and producers of chlorine, bleaching powder and sulphate of ammonia, were all subjected to standards and required to apply best practicable means. The standard fixed for other gases was not onerous – four grains of sulphuric anhydride per cubic foot. Control of the inspectorate passed from the Board of Trade to the Local Government Board. In addition to the industries regulated for the first time, the inspectors were to register salt and cement works, and to enquire whether emissions could be controlled at reasonable expense. Salt works, except those using harmless rock salt, were brought under control in 1892, but it was only in 1935 that cement works had standards imposed on them. In cement manufacture the main problem was not acid or smells, but dust, and not until the 1920s did electrostatic deposition offer a feasible technique for its control. Willesden power station had precipitators by 1929, a Kentish cement works applied the process in 1933, and from 1935 all new cement works were obliged to do likewise.

The number of works and processes over which the inspectors had control, or at least a watching brief, expanded rapidly. When Angus Smith compiled the first register in 1864 he had 84 alkali works on his books. By the end of the century when alkali was only one, though the largest, of the industries there were 1,000 works and 1,500 processes registered. In 1974 there were 2,000 works and 3,000 processes on the register. For the most part, the expansion of the workload of the inspectorate arose from extensions of the Alkali Acts, rather than from extra firms entering an already regulated trade. In 1892, for example, alkali waste works (a potential source of pollution by sulphuretted hydrogen), tar works, and carbonising works were among the trades subjected to the Alkali Acts. In 1906, oil refineries and tinplate works came within the net. The widest extension of the duties of the inspectorate came with the Clean Air Act of 1956. Unofficially, inspectors from Angus Smith onwards had concerned themselves with smoke, but they had no power in the matter until 1958 when industrial smoke ceased to be the responsibility of local authorities and came to the alkali inspectorate instead.[26]

By then domestic fires were the main source of smoke in the atmosphere. The improvement in London's record of winter sunshine between 1880 and 1920 did not continue in the inter-war years and was not resumed until the later 1950s. In the summer months too the standards reached by 1916–20 were not bettered for many years.

Table 2.3 Winter sunshine in central London and Kew

(London as percentage of Kew)

	%		%
1881–85	20	1940–50	55–60
1916–20	53	1950–55	26
1930–35	52	1955–60	68

Source: N Shaw and J S Owens, *The smoke problem of great cities* (1925), pp 65–6.

Fog, which includes natural fog, was less frequent in central London than at Kew or Heathrow. This surprising fact can be explained by the higher temperatures at night in the centre of large cities; it does not mean that the incidence of pea-soupers or smog was lower in central London, for that kind of fog arises only in cold weather with warmer air overlying the cold, contrary to the usual order of things. Murkiness, a smoky atmosphere in which visibility falls below a kilometre, occurred for over 900 hours a year in central London in the decade 1947–56, for only 600 hours at Kew, and for 500 hours in the country. Other great cities – Manchester, Glasgow, Leeds – also suffered like London from a considerable loss of winter sunshine as late as 1939–45. Glasgow and Manchester fared badly in the summer months as well. A hundred years after the first tentative attempts had been made to lessen the nuisance from smoke, the problem remained serious. Town halls, cathedrals and railway stations, especially in northern cities, presented a uniformly black face to the world, and nobody thought it worthwhile to clean them while soot and dust continued to rain down from the skies. Clothes got dirtier and furniture dustier, lungs more bronchitic, and cheeks paler in the big industrial cities. In April 1950 Dr. Barnet Stross, MP for Stoke-on-Trent, wrote to Aneurin Bevan, to ask what was the Minister's policy on atmospheric pollution; in Stoke, industrial haze occurred forty times as often as in Coventry. Radical though he was, the Minister tamely signed the reply prepared for him by his civil servants. It emphasised the practical difficulties and looked for cooperation between local authorities, industry, alkali inspectors, and regional officers of the Ministry of Fuel and Power:

I shall, of course, watch with interest the operation of recent local act powers relating to industrial furnaces and to smokeless zones, to see if they could usefully be adopted in the general law: but new powers are of limited use while practical difficulties remain. (PRO/HLG 55/58.)[27]

Only a severe shock would disturb the fatalism with which the public as well as government approached the evil of atmospheric pollution. The shock was not long in coming.

REFERENCES

1. N Shaw and J S Owens, *The smoke problem of great cities* (1925), pp 64–6, 91.
2. B R Mitchell and P Deane, *Abstract of British historical statistics* (1962), p 115 ff; S Pollard, 'A new estimate of British coal production, 1750–1850' (*EcHR* XXXIII, 1980), pp 213–35; *Annual Abstract of Statistics* v.d. [*AAS*].
3. Coal Commission, *Report of Committee E*, p 205 (PP 1871 XVIII); r.c. on coal supplies, *Final report*, p 17 (PP 1905 XVI); *Statistical Digest of the War* (HMSO, 1951) p 77; *AAS, 1961.*
4. E Gibbon, *Memoirs of my life*, ed G A Bonnard (1966), p 70.
5. [J Percy], 'Coal and smoke' (*Quarterly Review*, 1866), v. 119, p 451; J R von Wagner, *A handbook of chemical technology*, ed and trans W Crookes, (1872), p 733.
6. Smoke abatement committee, *Report of the committee* (London 1882), pp 173–4.
7. R.c. on coal supplies, *Evidence* (PP 1904 XXIII ii) qq 1327–85, 13321.
8. National Smoke Abatement Society [NSAS], *Proceedings of the annual conference, Glasgow 1953*, p 53.
9. Shaw and Owens, *Smoke problem*, pp 65–6; personal observation.
10. S.c. on smoke prevention, *Evidence* (PP 1845 XIII) qq 672, 676, 743–5, 1051; for the Bowling Ironworks, see British Association for the Advancement of Science, *Report for 1873* (1874), pp 220–1; A E Musson, 'Industrial motive power in the United Kingdom 1800–70' (*EcHR* XXIX, 1976), pp 432–4.
11. Musson, 'Industrial motive power', p 423; J W Kanefsky, 'Motive power in British industry and the accuracy of the 1870 Factory Return' (*EcHR* XXXII, 1979), p 374; Coal Commission, *Evidence*, qu 1167 (PP 1871 XVII); British Association for the Advancement of Science, *Report for 1872*, p 237; [Beaver] Committee on air pollution, *Interim report*, p 25 (PP 1953–4 VIII).
12. W H Booth and J B C Renshaw, *Smoke prevention and fuel economy* (1911), p 60; r.c. on rivers pollution, vol 2, *Evidence* (PP 1867 XXXIII), qq 10630–1; s.c. on smoke prevention (PP 1845 XIII) qu 462; Manchester Steam Users' Association, *A sketch of the foundation and of*

the past fifty year's activity (Manchester 1905), p 44; [Beaver] Committee, *Interim report*, p 491.

13. S.c. on smoke nuisance, *Evidence*, qq 135–41, 441 (PP 1843 VII); r.c. on rivers pollution, *Third report; evidence*, qq 11858 ff, 11967 ff (PP 1867 XXXIII); *29th annual report on alkali works*, pp 35–43 (PP 1893–4 XVI); r.c. on coal supplies, *Evidence*, qq 14972–7 (PP 1904 XXIII); Babcock & Wilcox Ltd, *Steam: its generation and use* (6th British edn, 1907), p 69.
14. W Fairbairn, *Iron* (Edinburgh, 1861), pp 82–5; r.c. on coal supplies, *Evidence*, qq 14710, 14857, 15597; r.c. on noxious vapours, *Evidence*, qu 11149 (PP 1878 XLIV); National Coal Board, *Report for 1947*, p 77 (PP 1947–8 X); W H Holmes, 'The pottery industry's contribution to clean air' (NSCA, *Clean air conference Eastbourne 1969*), pp 21–2, 28.
15. House of Lords, s.c. on noxious vapours, *Evidence*, qq 1751 ff, 1781 (PP 1862 XIV); r.c. on noxious vapours (1878), qq 2265 ff, 3175 ff, 3199, 3210, 3237, 3240; D W F Hardie, *A history of the chemical industry in Widnes* (Liverpool, 1950), pp 34–5.
16. T S Willan, *Navigation of the River Weaver in the eighteenth century* (Chetham Society, Manchester, 1951), esp. pp 208–28; T C Barker and J R Harris, *A Merseyside town in the industrial revolution: St Helens 1750–1900* (Liverpool, 1954), pp 4–6, 355; A and N Clow, *The chemical revolution* (1952), pp 86–7, 100–1, 275; *111th [and last] report on alkali works* (HMSO, 1975), p 19.
17. S Jubb, *History of the shoddy trade* (1860), pp 26–8; W S B McLaren, *Spinning woollen and worsted* (1889), pp 48–9; H Burrows, *A history of the rag trade* (1956), p 39; *21st report on alkali works* (PP 1884–5 XV), pp 7–8, 52; *30th report* (PP 1894 XIX), pp 71–2; *99th report* (1963) p 17.
18. R.c. on noxious vapours (PP 1878 XLIV), qq 966–78, 9744, 9765 ff, 9830 ff.
19. Health of Towns Commission (PP 1845 XVIII), App. pp 130, 135, 139–41; Lords committee on noxious vapours (PP 1862 XIV), qq 601–8, 997 ff; *25th report on alkali works* (PP 1889 XVIII), p 11; r.c. on noxious vapours (PP 1878 XLIV), qq 53–6, 10899 ff.
20. A Redford and I S Russell, *The history of local government in Manchester* (1940) II, pp 408, 424 and n; r.c. on rivers pollution, *Evidence*, II (PP 1870 XL), qu 1; ibid., *Report*, pp 23–4; *Municipal yearbook for 1914*, pp 889 ff describes sanitary arrangements in many towns.
21. R.c. on rivers pollution, *Evidence* II, (PP 1867 XXXIII), qq 6656–8; *Municipal yearbook for 1914*; r.c. on noxious vapours (PP 1878 XLIV), qq 9048 ff, 10763 ff; W Robertson and C Porter, *Sanitary law and practice* (3rd edn 1912), pp 102–4.
22. C S Orwin and E H Whetham, *History of British agriculture, 1846–1914* (2nd edn Newton Abbot, 1971), pp 14, 18, 33–4, 148; F M L Thompson, 'Nineteenth-century horse sense' (*EcHR* XXIX, 1976), pp 60–81, esp. p 77.
23. *99th report on alkali works* (1963), pp 13–14; Lords committee on noxious vapours (PP 1862 XIV) qq 2009–89; r.c. on noxious vapours (PP 1878 XLIV), qq 8152 and 8383.
24. S.c. on nuisance from smoke (PP 1843 VII), on smoke prevention (PP 1845 XIII); *Smoke prohibition: a report by Sir Henry de la Bèche and Dr Lyon Playfair* (PP 1846 XLIII); 16 and 17 Vic c 128 (1853); *Parliamentary*

debates, 9 August 1853, col 1496; Returns relating to the laws against smoke (PP 1854–5 LIII, 1861 L, 1866 LX, 1884 LXIII).

25. *Parliamentary debates*, 9 May 1862, cols 1452–67, 27 July 1863, col 1439; s.c. on noxious vapours (PP 1862 XIV), *Report*, viii, qq 2034, 2480; E Ashby and M Anderson, *The politics of clean air* (Oxford 1981), p 53; r.c. on noxious vapours (PP 1878 XLIV), qq 328, 12256; *105th report on alkali works* (1969), p 14.
26. Ashby and Anderson, *The politics of clean air*, pp 101, 131–2; NSCA, *44th Annual Conference 1977*, p 5.
27. Shaw and Owens, *Smoke problem*, pp 65–6; T J Chandler, *Climate of London* (1965), pp 125, 212; Department of Scientific and Industrial Research [DSIR], *Atmospheric pollution 1939–45* (1949), p 61; PRO/HLG 55/58.

Overleaf:
Sheffield, a smoky city. The domestic fire contributed at least as much as heavy industry to the smoky atmosphere of cities like Sheffield. After the passing of the Clean Air Act in 1956, Sheffield was one of the first authorities that took energetic action to rid itself of smoke.

CHAPTER THREE

From Smoke Abatement to Global Warming

DEADLY FOG

In 1945 the Director of Fuel Research in the Department of Scientific and Industrial Research surveyed the problem of atmospheric pollution. He noted 'a gradual, but very definite improvement in many areas. There is much truth in the statement that the old type of London fog is almost unknown'. His optimism was misplaced. As peace returned the home consumption of coal rose and thin fog still hung over London for between 40 and 60 days a year. This fog was defined as giving visibility of less than a kilometre, and was not what the Director had in mind. He was thinking of the Dickensian 'London particular', the pea-souper or smog.* Records of this severe fog – acrid, sometimes evil-smelling, and always dense – go back in London to the seventeenth century. John Evelyn recorded one in his diary for December 1671. Twenty-five such fogs occurred in the eighteenth century, and there were no fewer than 14 in the first 40 years of the nineteenth century. There were some notable fogs, usually in association with exceptionally cold weather, between 1873 and 1892. The worst, measured by the extra mortality from bronchitis in the week after each fog, occurred in 1880, 1891 and 1892. In 1880, deaths from bronchitis were 130 per cent above normal, in 1891, 160 per cent and in 1892, 90 per cent. In the worst year, deaths from all causes

---

* 'Smog' was coined from the words smoke and fog as early as 1905, but did not become widely known until the great London fog of 1952. Nowadays it describes the photochemical pollution associated with Los Angeles and the motor-car.

were 90 per cent above average. In 1918 there was again fog, but the high mortality then was mostly attributable to the influenza epidemic. In November 1921 there were two days of fog and cold but no obvious extra deaths. In 1924, fog without cold again added little to average deaths. The cold and fog of 1935 caused a 40 per cent rise in deaths from bronchitis. On this evidence it would seem reasonable to hope that London fog was largely a thing of the past.

Unhappily, in 1948 the old pattern reasserted itself; a spell of cold foggy weather doubled the deaths from bronchitis and increased mortality in London by 30 per cent. Much worse was to come in 1952. A cold spell began in mid-November, but death rates did not increase until even colder weather, accompanied by dense fog, set in on 5 December. The temperature remained below freezing until 9 December when the fog began to lift. At times visibility was below ten yards, and for substantial periods it was below twenty yards. Fog invaded public buildings and on the night of 8 December the audience with balcony seats in the Royal Festival Hall could not see the stage. By the test of extra mortality, the fog of 1952 was four times as deadly as the fog of 1948 and considerably worse than the fog of 1891. While it lasted its effects were more severe than those of the cholera epidemic of 1866, in that deaths exceeded the norm by a larger margin. Only the influenza epidemic of 1918 among modern outbreaks of disease increased the death rate more sharply. It is fair to add that the cholera and Spanish flu epidemics lasted much longer than the 1952 fog. And none of these visitations can be compared with the toll exacted by bubonic plague in 1665.[1]

On the medical evidence the 1952 fog was far worse than that of 1948. When the quantity of pollutants in the air was measured, the difference narrowed, for by that test the 1952 fog was about half as bad again. About ten times the quantity of pollutants normal for the time of year was found in the air in December 1952. If more than one part per million of sulphur dioxide is present in air (2860 micrograms per cubic metre), breathing may be affected. Much higher levels were found. Over a 24-hour period concentrations of 1.34 ppm are known to have occurred and it is suspected that much higher concentrations existed but went unrecorded. (In the smog of 1962 even higher concentrations of 1.98 ppm were measured over a 24-hour period.) Concentrations of smoke at 4.46 micrograms per cubic metre were well above what was needed for a severe fog. A normal winter concentration of smoke in a residential/ commercial area of inner London would have been 0.5 micrograms.

Smoke is not simply carbon dust, but contains tar and soot. Not surprisingly, an oily greasy film reappeared in London houses during the fog within an hour of being wiped away.[2]

SMOKE ABATEMENT

The fog of 1952 was a sad misfortune for London, but a blessing in disguise for the cause of clean air. A comparable fog in Manchester, Glasgow or Sheffield would undoubtedly have attracted less attention and done little to stir city dwellers and Parliament from their habitual apathy. Smoke abatement societies had existed continuously since the 1880s, and the impetus behind them went back further still. The Smoke Abatement Committee that organised the exhibition at South Kensington in 1881–2 was a London body that drew its inspiration from two earlier societies: the National Health Society founded in 1873 by the medical journalist Ernest Hart, and the Kyrle Society founded in 1877 by Miranda and Octavia Hill 'to bring beauty home to the people'. Ernest Hart chaired the new committee and among its members were: Edwin Chadwick and Lyon Playfair representing the heroic age of the public health movement; Lord Aberdare and Edward Frankland, the one chairman of the recent royal commission on noxious vapours, the other a distinguished chemist and member of the rivers pollution commission; Sir Frederick Leighton, PRA; and Sir Hussey Vivian, a Swansea copper-smelter. After a successful run at South Kensington, the exhibition moved to Manchester under the auspices of a local committee that included the Earl of Derby, the Mayor of Manchester, local MPs (among them John Bright), C P Scott, and a manufacturer of municipal refuse incinerators T C Horsfall. The exhibition in Manchester stayed open for a month and attracted over 30,000 visitors; 'a large proportion of the admissions at 3d. and 6d. being of the working class'. (Smoke abatement committee, 1882, p 144.) This success demonstrated a certain degree of public interest in the question of clean air. But the London society, renamed the National Smoke Abatement Institution under the patronage of the Duke of Westminster, did not remain long in existence, and in 1898 a new body, the Coal Smoke Abatement Society, was set up. It had a continuous, but not very influential, existence until 1924 when it amalgamated with various provincial societies to form the National Smoke Abatement

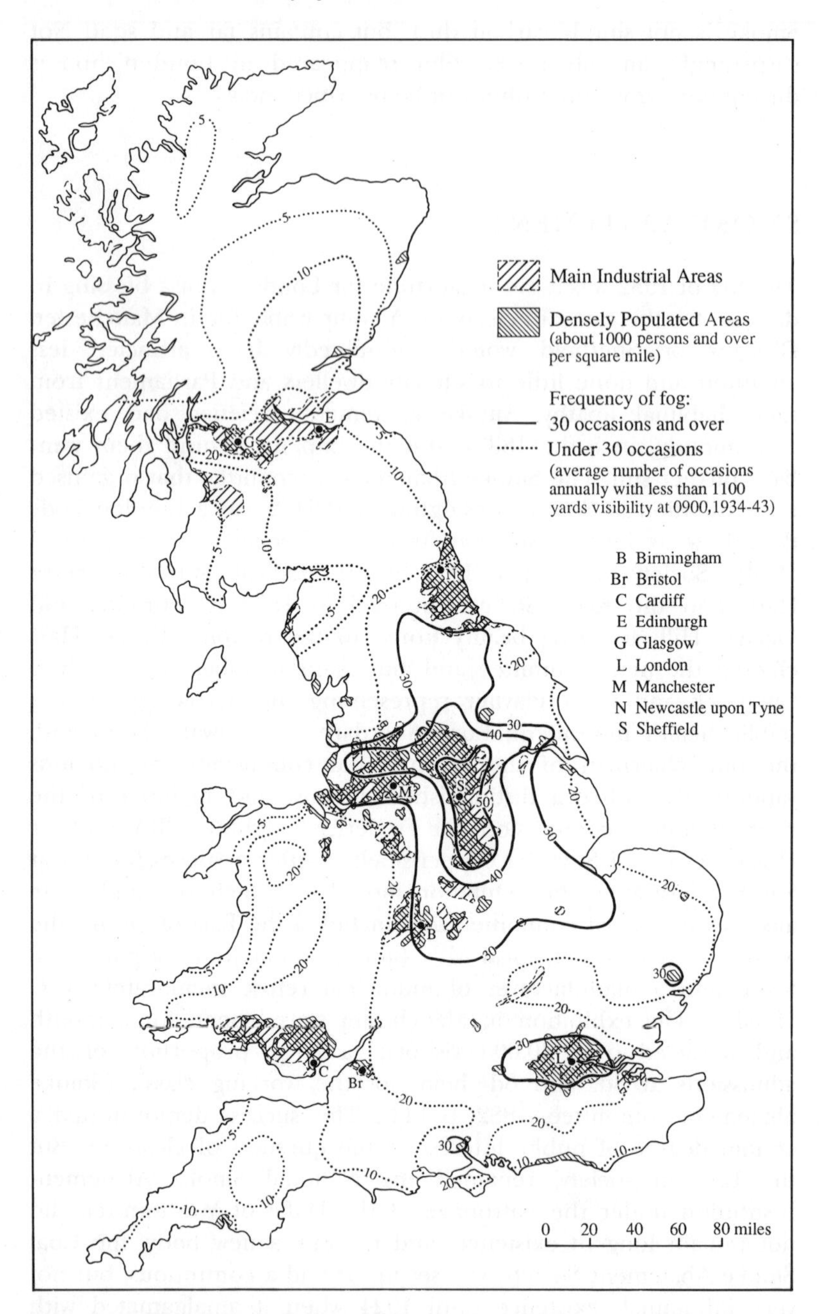

Map 1: Air pollution in Britain at 9a.m. (Source: *The Climatological Atlas of the British Isles,* HMSO, 1952)

Society. The term 'smoke abatement' shows the limited aims and small hopes of these various bodies. It was not until 1957 that the NSAS felt bold enough to change its name to the National Society for Clean Air, thus proclaiming a much more ambitious goal than Ernest Hart and Octavia Hill would have thought feasible.[3]

Visionaries had from time to time proclaimed the possibility of eliminating smoke. In 1843 a London solicitor put forward a proposal to make the burning of Welsh anthracite compulsory. Early Victorian England was hardly prepared for such a drastic interference with a man's hearth and home, even if supplies of anthracite had been equal to the task, which they were not. It would have been rather more prudent to advocate the use of coke. The editor of Evelyn's *Fumifugium* made a plea in 1772 for a method of charring (coking) coal so as to divest it of its smoke. In 1880 an alkali inspector demonstrated the possibilities by changing his heating and cooking arrangements; using coke instead of coal he saved money and eliminated smoke as well. Gasworks found a ready domestic market for their surplus coke but supplies were small compared to the market for raw coal. For example, in 1907 less than eight million tons of gas coke were produced at a time when the domestic demand for raw coal exceeded 30 million tons. In 1950 the gas industry made 12 million tons of coke* against domestic coal consumption of 40 million tons. The hard metallurgical coke produced by collieries and the iron and steel industry was unsuitable for use in the home. Proposals to eliminate the domestic use of raw coal by producing an extra 40 million tons of coke therefore fell on deaf ears as being improbable and unrealistic.[4]

By the end of the Second World War the contribution of the domestic fire to smoke and fog was widely understood even if the remedy remained unsure. A hundred years earlier, de la Bèche and Lyon Playfair had pointed out that in London the domestic fire made a lot of the capital's smoke. In factory towns it was clear at least to Angus Smith that industry was the principal offender. At 16 houses to the acre and five people to a house, perhaps 80 tons of coal would be burnt in a residential district; a factory covering an acre would burn 5,000 tons. No doubt there were more acres of housing than of factories, but factory smoke was a conspicuous evil – and one that could be readily tackled. The considerable success of

* Only a small proportion of the gas coke made was sold to domestic consumers at any time.

the effort to reduce industrial smoke is evident from estimates in the Beaver Report on Air Pollution in 1953. Homes burnt less than a fifth of the country's coal in 1952 but produced more than two-fifths of the smoke. Their output of sulphur dioxide was in proportion to coal consumed, since only a couple of power stations practised desulphurisation of flue gases.

While industry ignored the sulphur problem, it had made much progress in smoke prevention – hence the disproportionate amount of smoke from the unreformed domestic fire. It was fitting that Manchester should take the first steps towards tackling the problem of the domestic fire. Long notorious as the centre of a large area blackened by smoke, Manchester had seen an early ineffectual attempt to protest against atmospheric pollution – the Manchester Association for the Prevention of Smoke, founded by the Vicar of Rochdale in 1842. The Steam Users' Association of 1855 did practical good not only by its work against boiler explosions, but by giving members advice about smoke prevention. The smoke abatement exhibition of 1882 testified to the continuing desire to improve the quality of the air. Between the wars, Manchester built a large number of council houses, but installed conventional grates except for some 'all-electric' houses on one estate in the district of Burnage. Its only other practical contribution to cleaner air came from the dispersion of new estates in the suburbs.

THE SMOKELESS ZONE

The startling idea of smokeless zones originated in Manchester in 1935; a survey was undertaken three years later and in 1946 a local act gave the council the necessary power to establish smokeless zones. By 1953, 17 other authorities had followed this example and taken powers to create smokeless zones, though only Manchester and Coventry had put schemes into effect. The City of London, a little later in the field, had an easier task. There were few private residents, many of the firms with premises in the City (like the Corporation itself) were wealthy, and all suffered great inconvenience when fog brought London's buses and railways to a standstill. By the end of 1955 the whole of the City had become a smokeless zone in a remarkably swift response to the challenge of fog. Popular resistance to interference with the domestic fire was breaking down, as Coventry had found in 1948. In a referendum, ratepayers voted five to two in favour of smokeless zones, though perhaps without realising quite what they were committing themselves to. The idea of the smokeless zone was therefore to hand when a full-scale public enquiry was held into the London fog of 1952. The immediate reaction of the Ministry of Health to the disaster had been to set up internal committees of enquiry, but in the face of parliamentary protest a public enquiry chaired by Sir Hugh Beaver, a prominent industrialist, was appointed in May 1953. Working with unusual speed Sir Hugh's committee produced an interim report in six months and a final report a year later. Laying down the principle that clean air was as important as clean drinking water, and affirming that it would cost less to restore clean air than to allow pollution to continue, the committee recommended that in all heavily populated areas smoke should be reduced by 80 per cent within 15 years:

> It should be made the declared national policy to secure clean air. . . . Air pollution is a social and economic evil of the first magnitude. It not only does untold harm to human health and happiness, it is also a prodigal waste of natural resources. ([Beaver] (interim) *Report* pp 6, 11.)

The committee variously estimated the measurable waste at between £150 million and £250 million, about one or one-and-a-half per cent of the national income. Unburnt fuel, damage to stone and metal work and to furniture and clothes, the extra cost of lighting and soap, and the losses caused by delays to traffic during fog accounted for this formidable if somewhat speculative total. No estimate was made of the money cost of ill-health and death, perhaps wisely. For not only would it be difficult to estimate the value of labour lost, but also difficult to disentangle the many other agents of death and disease – tobacco smoke, unhealthy working conditions and overcrowded homes. Still less did the committee put a price on more distant evils such as acid rain, and the greenhouse effect. It was of course well known that huge quantities of sulphur dioxide were discharged into the air but there was no cheap or easy way to curb emissions. The threat to fish and forest from acid rain did not become apparent for another twenty years. The few scientists who knew of the possibility of global warming did not influence the Beaver Committee's thinking at all – perhaps they did not even try, since leading British meteorologists were then looking forward glumly to a new ice age, or at least to cooler weather than in the first half of the century. The greenhouse effect did not impinge on politics for another thirty years and more.

THE CLEAN AIR ACT

The immediate problems were formidable enough. Industrial smoke had been declining, not so much because of the pressure exerted by local authorities under the public health acts, as out of the self-interest of firms when cleaner fuels like electricity, or more efficient furnaces and boilers were installed. If industrial smoke was to be got rid of, which was technically feasible and not unduly expensive, the expert supervision of the alkali inspectors was appropriate. Their kid-glove methods were well known, and prosecutions occurred even more rarely than under Angus Smith, yet the inspectorate had some solid achievements to its credit. Not the least of these was that industry was willing to listen to its advice, but distrusted the officials of local authorities. The Beaver Committee therefore recommended that the alkali inspectors should assume responsibility for industrial smoke in certain industries such as iron and steel and the potteries. Industrial grit and dust, increasingly troublesome, were already within the inspectorate's terms of reference in certain industries. Local authorities would lose some little-exercised powers over industry but stood to gain new responsibilities over domestic fires. For the committee proposed that local authorities should be empowered to declare smokeless zones and to subsidise the change-over from smoky to smoke-free appliances. Government's initial response to these radical proposals was cautiously expressed in time-honoured and ominously negative phrases: 'Far reaching proposals . . . a number of quite complex technical problems . . . considerable sums of money . . . reserving their position in regard to individual proposals'. Nevertheless, the government accepted the report in principle and soon prepared legislation. The Clean Air Bill had its second reading in November 1955 and became law the following year. Few reports into a social evil have led to such swift action.[5]

The provisions of the act followed closely the recommendations of the Beaver Report. Sceptics feared that the prohibition of 'dark smoke' would have no effect since various defences were allowed such as breakdown of equipment, starting up a boiler or furnace from cold, or the use of unsuitable fuel. Industry was also given seven years' grace in which to re-equip with appliances which did not emit smoke. In the result, these concessions did not seriously weaken the force of the Clean Air Act since most large firms were willing to comply and it was difficult to put forward a successful defence based on unsuitable fuel, when with good practice almost any fuel would burn without smoke in an industrial boiler. Within ten years of the Clean Air Act, industry had reduced its smoke emissions by 74 per cent, so that it was domestic smoke that did most to pollute the air. By 1965 industry, once the worst offender, made only about one-fifth of the smoke in the atmosphere. Persuading householders to abandon their old ways was a much slower process.

Industry's other solvable problem was grit. Grit emissions had always been troublesome, for obvious reasons, at cement works. A new industry in the early nineteenth century, its output grew rapidly, reaching 7 million tons in the 1930s, 10 million tons by 1951 and a peak of 20 million tons in 1973. Cement first came under the control of the alkali inspectors in 1935. It was laid down that new plant had to include electrostatic precipitators and to limit emissions to 0.5 grains per cubic foot (cu ft). By 1968, when the permitted emissions had been reduced to 0.1 grains, only 12 out of 123 kilns were still without precipitators. Thus, despite a nearly threefold increase in output, emissions were lower than when control was first applied. The cost was considerable but not excessive. The capital cost for a new 12-kiln works amounted in 1968 to £158,000; in an old works the expense was three times as much. In a new works the filters added 7½d. (3.125p) to the price of a ton of cement. The running costs amounted annually to about a third of the capital cost. In all, the price paid to get rid of a serious nuisance cannot be regarded as high.[6]

Similar problems arose at steelworks with the new technique of sintering powdered ore before smelting. The process if uncontrolled threatened to shower the neighbourhood with a fine powder of iron oxide. In 1958, the year in which the alkali inspectors assumed responsibility for grit emissions from steel works, 130,000 tons of sinter dust escaped from a throughput of 8 million tons; by 1974 emissions were down to a third of the 1958 figure, although throughput had risen to 20 million tons. At power stations best practice required the crushing of coal before firing. The pulverised fuel undoubtedly burnt more efficiently; it also released huge quantities of fine dust – mostly ash – into the air. In

1958 when the stations were burning 43 million tons of coal, some precautions had already been taken in the form of high chimneys and grit arrestors. Yet estimated dust emissions amounted to a million tons, at the high rate of 0.8 grains per cu ft. By 1974 coal consumption had risen by a third, yet emissions were reduced to less than 200,000 tons at a rate of 0.1 grains per cu ft. To achieve these results heavy expenditure was undertaken, not too unwillingly, by prosperous industries untroubled by pressure to maximise profits at all costs. In the three years after the alkali inspectors took responsibility for steel works in 1958, British Steel spent £25 million on compliance at existing works. At the new Spencer Works anti-pollution costs amounted to four per cent of the entire cost of the project. Not all of the cost was unprofitable to the company since blast-furnace gases and iron oxide dust had substantial value. At a large power station costing £70 million in 1970 dust-removal equipment accounted for £3 million; a similar rate of expenditure to that at the Spencer Works. Dust was the only pollutant against which the electricity industry took any new precautions at that time; smoke was almost unknown in their modern boilers, and the removal of sulphur dioxide was costly and cumbersome. Until concern grew about acid rain, the pre-war experiments with desulphurisation at Fulham and Battersea power stations were not repeated, except at Bankside.[7]

COMPETITION FOR COAL

The Beaver Committee was confident that smoke from domestic fires could be eliminated. Its confidence was based on two assumptions: that adequate supplies of coke could be made available; and that conversion from open fires to closed stoves could be carried out cheaply. The first of these forecasts proved to be wrong. In the early 1950s the gas industry sold about 12 million tons of coke a year, but less than a quarter was bought by householders. Anthracite, hard coke, and patent fuels like Coalite and Phurnacite contributed another 2 million tons for household consumption, so that rather more than five million tons of smokeless fuel was available, or one-seventh of domestic coal consumption. Ten years later rather less coke was being produced by the gas industry. The demand for gas was growing rapidly, but was being met by the increasing use of oil for enrichment of coal gas, and by the use of natural gas or methane. Import of liquefied natural gas from North Africa began in 1964 and in 1965 the first gas strike was made in the North Sea. In 1967, North Sea gas began to be sold in Britain. Output of gas coke fell rapidly and ended in 1975. Householders complying with the requirements of the clean air policy rather suddenly found that a preferred fuel had disappeared. The capital costs of converting to coke-burning were low – less than half the cost of a gas fire or of a fixed-flue paraffin heater. The running costs were also low, with only paraffin and hard coke costing less.

It was therefore a matter of some official concern that supplies of coke were falling just as the attack on domestic smoke was getting under way. But the public was more adaptable than the official mind had allowed. The price of oil was falling relative to coal and coke, and gas and electricity were clean and convenient fuels. Fireplaces and chimneys were often omitted from the design of new houses and especially new flats in the 1960s. Ripping out old fireplaces became second nature to the do-it-yourself enthusiast intent on modernising his home, and the television set replaced the hearth as the central feature of the living room. Not content with traditional standards of domestic discomfort, increasing numbers of householders installed central heating, a device almost unknown in British houses before the 1950s. In theory, bituminous coal and its attendant smoke could have been burnt in solid-fuel central heating boilers where the law allowed; in practice coke or anthracite was almost invariably the chosen fuel if solid fuel was chosen at all. Throughout the sixties oil gained rapidly in popularity and the sale of solid-fuel central heating boilers fell to low levels.* Central heating by means of electric night-storage radiators was also popular for a time. In the 1970s when oil became expensive, gas-fired central heating came to the fore and is now predominant. In 1991 part or full central heating was enjoyed by 84 per cent of households (25 per cent in 1969, 50 per cent in 1977). Roy Brooks, the estate agent who amused his readers by running down the houses he had for sale, dealt in property of above average value. Yet

* When in 1970 the author chose an anthracite instead of an oil-fired system, the workmen who were installing it openly regarded the decision as eccentric and secretly as mad. The author's only defence – not a bad one – was that supplies of anthracite were more secure than those of Middle East oil. An expert might have added that a coal-fired system was simpler and less liable to breakdown.

in the late 1950s only one client in fifteen had the benefit of central heating. New council houses and flats built in the 1950s and 1960s were often heated by electricity, at the dictate of architects and housing committees, without much regard for the tenants' ability to pay. Like central heating in the private sector, electric fires in council flats were developments largely unforeseen by the Beaver Committee. Both made substantial contributions to ridding the atmosphere of smoke, sometimes in conjunction with smokeless zones, sometimes not.[8]

The Beaver Committee was especially anxious to impose smokeless zones in so-called 'black areas', densely populated cities and industrial districts where atmospheric pollution was at its worst. Local authorities responded with varying degrees of vigour to the opportunities afforded by the Clean Air Act. By 1963 only 14 per cent of premises in 'black areas' were subject to smoke control orders. Four years later, and eleven years after the passage of the Clean Air Act, the percentage of premises subject to the orders had risen to forty. Enthusiasts for smoke abatement were not impressed, but the historian can afford to be more charitable: the evil of atmospheric pollution had diminished so little in the first half of the twentieth century that the marked improvement after the London fog of 1952 was all the more striking.

In London itself, where smoke control orders were thick on the ground, most of the central area was subject to control by 1967 and two-thirds of premises throughout the area of the GLC (Greater London Council) were in smokeless zones. In the smokiest parts of the capital – North Kensington, the Strand and the City – winter concentrations of smoke averaged less than a third of what they had been on a typical winter's day in the early 1950s. By 1973 the amount of smoke in the air of central London had been reduced to a fifth of the level considered normal twenty years before, and there was little difference in the amount of winter sunshine enjoyed at Kew and in the centre of town. Sheffield was another city where smoke abatement was energetically pursued. In the winter of 1955–6 in the Don Valley where steelworks and dense housing were to be found together, and to the south-west of the city centre in an area of dense housing, concentrations of smoke occurred that were higher than in the great London fog of 1952. In the city centre where there was little housing and less heavy industry, the pollution was less severe, but at 2,900 micrograms per cubic metre was not to be disregarded. The centre came under smoke control orders in 1959 and in 1971 the Don Valley, the last area of the city to be tackled, came within the orders. Average winter concentrations of smoke fell sharply everywhere. Even in the Don Valley smoke concentrations had fallen by three-quarters in the winter of 1970–1. Sheffield, formerly notorious for its smoke, enjoyed pretty clear skies by the mid-1970s, and an atmosphere as clean as London's.

What Sheffield had done could have been done in other northern cities too, but they were slower to use their powers. In Manchester and Liverpool, for example, the air was decidedly smokier in the late sixties than in London or Sheffield. Smaller towns in Lancashire – St Helens and Runcorn – had somewhat lower levels of smoke. Everywhere skies were clearer than in the past because domestic coal consumption was falling, smoke control zones or not. The skies were of course clearest where smoke control existed, and smoke control provided an extra incentive to choose an alternative fuel. It was noticeable that domestic coal consumption varied widely from region to region in 1970. In London other fuels had almost entirely replaced coal, annual consumption of coal having fallen to less than a hundredweight (cwt) per head. Northern England and Wales at the other extreme still consumed over ten hundredweight. A close correlation naturally existed between coal consumption and smoke particles measured in the air, except in South Wales where the coal was virtually smokeless. The national survey of air pollution, an important source of information and guide to policy, came to an end in 1982. Confining itself to pollution by smoke and sulphur dioxide, it no longer appeared to a parsimonious government to give value for money. Domestic consumption per head per annum had fallen to less than two cwt of bituminous coal, about a seventh of the pre-war level, and smoke had virtually disappeared from the skies. The problem of sulphur dioxide remained, though on a reduced scale. Public attention was turning to new-fangled pollutants, less obvious but to many people quite as worrying as old-fashioned smoke and fog.[9]

INVISIBLE POLLUTANTS

Sulphur dioxide can be regarded as a new-fangled worry only in the sense that an unscientific public was unaware of its prevalence and dangers until quite recently. Experts like Angus Smith well knew that the tonnage of sulphur dioxide poured into the air in 1880 greatly exceeded the tonnage of hydrochloric acid. Every ton of coal contains sulphur in varying proportions, averaging about one and a half per cent. Some remained in the ash after combustion; most escaped into the air and being relatively insoluble in water, was widely dispersed. Nevertheless, concentrations were high in large towns, corroding metal and stone. (The acid atmosphere also inhibited the growth of lichens, and encouraged rose-growers, since the black-spot fungus could not live in the acidic air of great towns, and has only returned in recent years as pollution has diminished.) Fuel oil contains much more sulphur than coal – three or four per cent against one or two per cent in coal. The displacement of coal by oil that proceeded rapidly between 1945 and 1973 therefore threatened to exacerbate the problem. It has proved feasible however to extract sulphur from oil in the refining process; and so the rising consumption of oil has not added as much sulphur dioxide to the atmosphere as might have been expected. All figures for emissions of sulphur dioxide make assumptions about the average sulphur content of fuel and about its retention in ash. Table 3.1, following, must therefore be taken with a pinch of salt as a rough guide to probabilities, rather than as precise measurement:

Table 3.1 Sulphur dioxide emissions in Britain 1800–1986

(million tons)			
1800	0.25	1965	5.80
1870	2.50	1968	6.00
1900	4.00	1974	5.00
1939	4.00	1977	5.00
1951	5.00	1986	3.70

Higher emissions should not be confused with higher levels of acid gases in the local air. During smog, dangerously high concentrations did occur. In London in 1952, levels as high as 2,700 to 3,800 micrograms per cubic metre were recorded; even higher levels (5,660 mg) were recorded in London in 1962 when another dense fog occurred. In a London smog in 1975, the pollution was much less serious, reaching a maximum of 1,200 mg per cu m. Any level above 500 mg is likely to distress sufferers from respiratory problems and such levels were exceeded in London, the Midlands and the North on 'very many days' in the 1960s. The national trend, apart from exceptional events like smog, was downward, thanks to higher chimneys and to the increasing use of natural gas.

Average concentrations of sulphur dioxide fell by 29 per cent in the 1960s to 118 micrograms and have since fallen much further.

In 1987 the electricity industry consumed not less than 30 per cent – over 90 million tons – of the supply of fossil fuels, including three-quarters of the supply of coal. Since power station chimneys range in height from some 200 ft to 850 ft it is not surprising that local concentrations of sulphur dioxide are considerably less than in the past. In practising one of the favourite remedies for pollution – dispersion – the electricity industry has been so thorough that much of Britain's sulphur dioxide is exported, without permission, to Scandinavia. The transfer must have been going on for a long time, but it was not until 1968 that the phenomenon of acid rain first came to light in Sweden, and was linked to the acidification of Swedish lakes and rivers. Acid rain has since been blamed for the death of forest trees as well, but the electricity industry has been unwilling to accept responsibility. In the subsequent and still continuing debate, hot dry summers, ozone poisoning and car exhausts have competed with sulphur dioxide for the doubtful fame of killing Europe's trees. It may be that the explanation lies elsewhere, perhaps with infection. Or that until recently nobody had bothered to count dead trees; like other living things, trees must be at some risk of premature death, and the question arises whether the proportion of trees dying is decisively higher now than in past centuries. Although sulphur in small quantities is essential to life there can be no doubt that excessive atmospheric doses are inadvisable. How much, if any, damage has been done by acid rain is still an open question, but even if none has been done to fish and trees, the damage to buildings and at high concentrations to human lungs, is undeniable.

The electricity industry has therefore been driven not only to sponsor research into acid rain, but to do more to extract sulphur from flue gases. Fulham and Battersea power stations, with difficulty and expense, intercepted most of the sulphur in flue gases in the 1930s. Suspended during the war, desulphurisation was reinstated afterwards, and was also practised in the oil-fired station at Bankside opposite St Paul's Cathedral. Experience with the process did not encourage the Central Electricity Generating Board (CEGB) to install similar equipment elsewhere. Not until 1987 did it agree to install a desulphurisation process using limestone in all its new coal-fired power stations, at a cost of £600 million. Some older stations with a capacity of 6,000 MW are also to be modified at the further cost of £170 million. Since research published by the industry continues to play down the dangers of acid rain the pressure of opinion both at home and abroad must have been severe to make it incur this considerable expense, apparently against its better judgement. With privatisation, the consumption of coal in power stations is falling, and it is uncertain how far flue-gas desulphurisation will now proceed.

GLOBAL WARMING

Acid rain is one of several modern environmental hazards discovered (the unkind might murmur 'invented') by scientists. Another and more recent one is global warming, a danger not widely appreciated before 1987. Meteorology is a study so complex, that despite a plethora of information, forecasting the weather even a week ahead is beyond the power of man or computer. It does not necessarily follow, however, that forecasts of global warming are unreliable because they are to be tested over a period of decades rather than days. Large events are often easier to predict than small events: nobody can tell within a wide margin of error how many people will die in road accidents tomorrow, but knowing one year's deaths it will be safe to predict that next year's will not normally deviate from that level by much more than five per cent. If meteorologists could accurately measure all the factors in the formula for global warming, and if their formula or model faithfully reflected reality, they could declare with some confidence the likely rise in world temperature by (say) 2050. Some facts are indisputable, if past and present techniques of measurement can be relied on: carbon dioxide is now present in the atmosphere in the proportion of 350 ppm, against 315 ppm in 1958 and only 270–280 ppm in 1850. It is estimated that only about half the carbon dioxide discharged into the atmosphere since 1945 remains there; the rest has been absorbed by the oceans and by vegetation. The future output of carbon dioxide is unknown though shrewd guesses can be made; the future absorptive capacity of sea and forest is quite uncertain. It is reasonable to suppose that chlorofluorocarbons (CFCs) will not long continue to damage the ozone layer since the international agreements of 1987 and 1990 should lessen their production. Of the other greenhouse gases, methane or natural gas is probably the most important. It is present in the air in minute quantities (0.65 ppm in the late eighteenth century, 1.7 ppm today), but it is increasing at the rate of one per cent a year, and will therefore double from present levels, unless checked, in some 70 years. As a greenhouse gas it is much more powerful than carbon dioxide and its increase therefore poses a considerable threat. There is already a little direct evidence for global warming. The world's average temperature has risen by about 0.5°C since the beginning of the century, though rather awkwardly for the theory, temperatures actually fell between 1950 and 1970. What is more menacing is the much larger rise in Alaskan temperature – up 2.2–3.9°C in the past 100 years. For a substantial warming in high latitudes brings with it the danger of melting ice and higher sea levels. Sea levels are said to have risen about 20cm between 1900 and 1970, though at a slow rate after 1940. Predictions for the future suggest a rise of 20–140 cm over the next few decades, if temperatures rise by from 1.5 to 4.5°C. The five hottest years since 1852 have all occurred in the 1980s – 1980, 1981, 1983, 1987, and 1988. The explanation may lie in natural causes not yet understood, but whether the greenhouse effect is to blame or not, ice is going to melt all the quicker as a result of the warm spell.[10]

Meteorology is neither an exact nor a monolithic science, and there is much that is doubtful in the theory of global warming. Two-thirds of the world's surface is covered by water and the influence of the oceans upon future climate is far from clear. Other uncertainties concern cloud cover: if temperatures rise, more water will be evaporated and there may be more cloud. Whether on balance the cloud will act as a blanket making the world warmer, or as a mirror reflecting solar heat back into space is a moot point, and computer simulations do not offer as sure an answer as a long run of meteorological observations. The consequences of deforestation are also uncertain. If a large part of the Amazonian rain forest is felled the bare earth of Brazil will reflect back much more heat into space than the trees do at present, so tending to cool the earth. At the same time fewer trees will probably lead to clearer skies and lower rainfall, with doubtful consequences for world temperatures.

The uncertainties surrounding the greenhouse effect extend still further. On a world view, higher temperatures may prove advantageous. The Nile Delta, Bangladesh, Holland, eastern England and many other low-lying populous regions will suffer disastrous flooding. On the other hand, it is an agreeable prospect that Mediterranean and other sub-tropical climates may extend to higher latitudes than at present, changing the crop pattern and moving the tree-line further north in Canada and Russia. It is at least possible that the world could support a much larger population, or a stable population at a higher level of comfort, after the greenhouse effect had occurred. Britain, however, with large areas of fertile and heavily peopled low-lying land would almost certainly lose much more from flooding than it would gain from a warmer climate. The principle of self-interest therefore suggests that no British government can afford to ignore the theory of global warming despite its many uncertainties.

NEW SOURCES OF ENERGY

Britain burns only about three per cent of the world's annual production of fossil fuels, so that if thc British Isles sank beneath the waves, the reduction in the greenhouse effect would not be large. Within the narrow limits of the possible, Britain can lessen its consumption of fossil fuel (or diminish the rate of increase) in two ways: by energy conservation,* and by the use of alternative sources of energy. New sources of energy do not quickly displace old, if new technology is called for. The triumph of the steam engine over wind and water power in Britain was scarcely complete before the end of the nineteenth century or nearly two centuries after the construction of Newcomen's first fire or atmospheric engine in 1712. Murdock first lit his house in Redruth by gas in 1792, but oil lamps and candles were still common at the end of Victoria's reign when gas-lighting in its turn faced a new rival, electricity. Given the relatively slow rate of capital accumulation in modern Britain it is not to be expected that new sources of energy will be developed quickly. Harnessing the tidal power of the Severn Bore, for example, has been under discussion for at least 70 years. In the expansive phase of reconstruction after the First World War, when large schemes were in the air, one such scheme was for a barrage across the Severn that would generate electricity equivalent to the combustion of 1.25 million tons of fuel in an efficient power station, or of twice as much fuel in small power stations or factories. The scheme was turned down, as was a more ambitious one proposed by the Bondi committee in 1981. Sir Frank Layfield considered the Bondi proposals as part of his wide-ranging public

* For energy conservation, see Chapter seven.

enquiry into the planned Sizewell B nuclear power station. On reasonably favourable assumptions, the Severn Barrage might yield a return of five per cent, he found, and supply five per cent of the electricity at 1987 levels of demand. Since the barrage could not be ready before the year 2000 and would not yield as high a return as Sizewell B, Sir Frank could not recommend it. That was a questionable judgement, for the likely costs of Sizewell B have since increased. The Severn Barrage remains a tantalising prospect and yet another report appeared in 1989. More ambitious than any of its predecessors, it assessed the capacity of the barrage as seven per cent of the electricity supplied at present by coal-fired power stations. As with all major civil engineering projects the estimated cost seems temptingly low. Construction of the barrage, if it takes place, will itself demand a great deal of conventional energy but if, as is forecast, the barrage has a life of a hundred years, the saving of fossil fuels will be considerable and the benefits to the environment correspondingly great, in the long run. More difficult problems are set by attempting to harness wave power off shore. The electricity industry has in a small way been sponsoring research into wave power for nearly twenty years. It has also encouraged experiments in the generation of electricity by wind power.[11] Neither wind nor waves seem likely to add much to energy supply in the next ten or twenty years. Geothermal power, or energy from hot rocks underground, is an even more distant prospect.

In the immediate future, government, if not the general public or the City, seems to regard nuclear power (if it can pay its way) as the best alternative to fossil fuels and the best antidote to the greenhouse effect. The history of nuclear power is a telling example of the difficulties that face a new technology. The earliest nuclear reactors were not expected to produce power that could compete on equal terms with a conventional power station. Their purposes were scientific and military. And it may be suspected that 'a cheap clean source of power' remains a more attractive slogan than 'give us more plutonium'. But until the archives are open this suspicion can be neither confirmed nor denied. Whether nuclear power has yet proved a cheap source of electricity is not a relevant consideration in a chapter about pollution. It is the hazards of radiation that are relevant and have long excited controversy. It seems to be the case that there is no safe dose of radiation: any exposure may cause mutations and cancer. By the same token it could be said that there is no safe way to cross a road, or to drink a glass of water. In every case the risk, if it is considered at all, has to be set against the advantages.

Where deaths can be directly compared in the power industries – among workers in nuclear power stations and in coal mines and coal-fired power stations – nuclear power turns out to be much safer. For every death among nuclear workers, there are seven, mostly among coal-miners, in the conventional coal-fired electricity industry. The report from which this information comes does not examine the environmental consequences of radiation and fog. There too the balance of advantage appears to lie with nuclear power. 87 per cent of the radiation occurs naturally; another 12 per cent comes from medical irradiation; and only about one per cent from nuclear weapons fall-out and nuclear power. The average Briton does not of course exist, except in statistical imagination, nor does he die of lung cancer or leukemia, but it may well be more dangerous to live in some parts of Devon or Cornwall than in the shadow of Sellafield or Dounreay. The risk of death or injury from the man-made radioactivity of nuclear power stations, their reprocessing plant and effluent discharges is peculiar to the areas in which the nuclear plant is located. There is good evidence of a higher incidence of leukemia near to nuclear installations, but it remains even there a rare disease. There is no marked difference between the health record and mortality of areas close to and distant from nuclear installations, and at present the risk of death from man-made radiation appears to be small. Yet there is no doubt that the fear of radiation is common and borders on dread. On a long view, the fear may not be unreasonable. Old-fashioned atmospheric pollution does not persist once the fires are put out. The pollutants fall to earth as dust or acid and are absorbed fairly quickly. Nature recycles most pollutants and makes them harmless within a short time; and there is a rough justice in the symmetry under which those who benefit from the warmth of the open fire and the convenience of the motor car also suffer from the smog. There is no such justice with radioactivity. Its benefits, like those of the coal fire, are momentary, but its waste products, unless safely stored, can deal out death for perhaps thousands of years to come. If nuclear power continues to be used, its waste products will accumulate until they are a much larger part of the total radiation dose than at present. If nuclear power is abandoned, future generations who get no benefit from it will remain at some risk from its waste products – a strange bequest to our children's children. It is this cumulative and persistent character that distinguishes nuclear pollution and makes it hated and feared.

Radioactivity is the most spectacular of the threats to clean air posed by modern scientific discovery. There have been many others, of which only a few will be considered here. One is the addition of lead to petrol, a practice that began over sixty years ago. Lead poisoning was known to the ancients: both Hippocrates and the elder Pliny described it. In modern times, lead poisoning was noted among English white-lead workers in 1678. In 1839 a French physician Tanquerel des Planches published a detailed account of 1,200 cases of 'plumbism' and by 1864 the factory acts were prescribing some limited precautions against it in English potteries. In 1883 an act was passed to regulate working conditions in white-lead manufacture and in 1892 Britain appointed its first medical inspector of factories, T M Legge, who did important work on lead poisoning. Greater care and new production techniques sharply reduced the number of cases of lead poisoning notified among workers from over 1,000 in 1900 to 55 in 1960. The quantities of lead added to petrol were small – about half a gram per litre – and for many years attracted little criticism. By the early 1970s lead additives began to come under public scrutiny as part of the heightened concern over environmental hazards. In 1971, before any controls had come into force, 7,300 tons of lead passed into the atmosphere from car exhausts. In 1981 the figure was slightly lower at 6,700 tons while in 1986 after fairly strict limits had come into force, the amount of lead released fell to 2,900 tons. Lead-free petrol was at that time a rarity, sold at only 200 garages. Substantially lower rates of taxation of lead-free petrol have since transformed its commercial prospects. Unleaded petrol in 1992 accounted for a half of all sales as more and more motorists had their cars converted, with no loss of efficiency, to run on the cheaper fuel. A government dedicated to the promotion of market forces might well feel gratified at this proof of British adaptability. Nor have interventionists any reason for displeasure since it was Nigel Lawson's decision to distort the market by lowering the tax that gave the impetus to the spread of unleaded petrol.[12]

The dangers from leaded petrol were not as acute as those facing workers in potteries or in paint manufacture. But they could have caused some brain damage to persons exposed to heavy traffic, especially to children. The risk was known long before action was taken, in accordance with the law of politics that reform depends on protest more than on the merits of the case. Unlike lead additives, there are other substances once thought harmless that now cause alarm as a result of bitter experience. Asbestos is a good example of a once innocent building material that now causes flats and offices to be evacuated and stripped of the offending material. A man-made substance that has a similar history is the gas, vinyl chloride monomer (vcm), a constituent of the plastic, polyvinyl chloride (pvc). At one time US doctors had considered using vcm as an anaesthetic, and manufacturers in Britain as elsewhere took no undue precautions against its escape. Losses of 50 lb of vcm for every ton of pvc made were not uncommon in the early 1970s and at six works the loss amounted to 88 lb a ton. The permissible limit of exposure in plastics factories was set by the alkali inspectorate at 200 ppm over an 8-hour day. At this point American research showed that the once-candidate anaesthetic caused liver cancer among plastics workers. Permissible limits of exposure in Britain were promptly reduced from 200 ppm to 25 ppm over the working day, with temporary maxima of 50 ppm. In view of the dangers, these do not seem to be particularly stringent rules. Outside pvc works, 90 per cent of air samples showed concentrations of 0.2 ppm or less with a maxima of 2 ppm.[13] Whether these low levels represented a threat to health was far from clear. It is not too fanciful to imagine a state of affairs where it will one day be possible to predict which new chemicals will prove harmful to man, beast, or vegetation, before they are marketed. In the present state of knowledge, control usually follows rather than precedes an unforeseen and unknowable disaster, and society seems willing to pay that price for the advantages that new substances offer. To ban them until they are proved harmless would burden enterprise with extra costs, and slow the rate of technical change. The preference seems to be for taking some risks with public health.

EFFECTS OF AIR POLLUTION ON HEALTH

It is not easy to establish how much damage atmospheric pollution, or any one adverse circumstance, does to health. The low death rates experienced in late twentieth-century Britain and the steady rise in life expectancy suggest that the danger from recent hazards has been small, and that even in Victorian times the population was surprisingly resistant to the perils of life. Each successive life-table drawn up by the Registrar General showed some reduction in mortality. The mid-Victorian years of cholera, heavy immigration from Ireland, and rapid uncontrolled growth of towns showed least improvement. Since the middle of this century the degenerative diseases of old age have predominated among the causes of death; it is no wonder therefore that further reductions in mortality have proceeded at a somewhat slower pace.

Table 3.2 Expectation of life at birth: England and Wales

	(in years)	
	Men	Women
1838–54	40	42
1871–80	41	45
1901–12	52	55
1930–32	59	63
1950–52	66	72
1984–86	72	78

In a catalogue of the world's ills, an expectation of life of forty years would not count as a major source of misery. A century-and-a-half ago, many countries had higher death rates than England and until recently the expectation of life in poor countries in Asia was much lower than in early Victorian England. Yet in comparison with modern Britain, the life-table for 1838–54 depicts a heavy loss of human life, some of it almost certainly attributable to atmospheric pollution. From the beginning of civil registration in 1837 until the 1950s the large towns of England and Wales experienced higher death rates than the countryside. On almost every other count towns looked more inviting than the 'idiocy of rural life'. Countrymen all over Europe from Ireland to Russia testified to this fact by escaping to them in an irresistible flood. There is good evidence that the English farm labourer's wages and diet were lower than those of town workers. Urban homes were as well built and no more overcrowded than those in the countryside; urban sanitary arrangements, at their worst in the 1830s and 1840s, improved somewhat thereafter. Sanitation and water supply remained primitive in the countryside for many years after clean running water and main drainage had become the urban norm. Until the provision of allotments became popular in the 1880s, it was unusual for farm labourers to have plots large enough to grow vegetables, let alone fruit, for their families. The diet of the working class was regularly deficient in fruit and vegetables, as was shown in the Board of Trade enquiry into urban family budgets in 1904. Most families spent about 2.5 per cent of the family income on fruit and vegetables other than potatoes – rather less than they spent on tea or sugar. Paradoxically, fruit and vegetables tend to be cheaper and fresher in towns than in villages; it is therefore likely that the farm labourer – unless he had a garden or allotment – ate even fewer carrots, cabbages and apples than the town worker. Yet the farm labourer and his children were and remained, the healthiest members of the working class, although, until the last quarter of the nineteenth century, their only conceivable advantages were sunshine and fresh air.

Dr William Farr* was the medical statistician who first investigated occupational mortality on a quantitative basis. It had long been clear from the annual reports of the Registrar-General that mortality in towns was substantially higher than in the country. Farr went further and in 1865 asserted a close correlation between density of population and mortality. He also made his second and not entirely satisfactory contribution to occupational-mortality statistics. He returned to this subject in 1875 and by separating occupational groups more carefully than before, was able to show the huge disparity of death rates. By allowing for age at death, Farr was able to show that the clergy, gamekeepers and farm labourers all enjoyed long life; over half the deaths among the clergy and farm labourers for example, occurred after the age of 65. In woollen manufacture, still partly a handicraft and semi-rural trade, almost as many lived into old age. But among cotton workers only a quarter, and among coal miners only a fifth lived to their sixty-fifth birthday. When Farr's successor repeated this study for the years 1900–2, open-air work together with the church, again offered the best hope of long life: gardener, gamekeeper, farmer, railway-engine driver or stoker, and farm labourer were the healthiest occupations after clergyman.[14]

If clean air promoted long life, diseases of the lungs should have carried off many townsfolk. That was indeed the case, and the link between smoke and illness had been noted at the end of the seventeenth century by Gregory King, one of the founders of English demography. When reasonably reliable mortality statistics

* Farr (1807–83) was of humble origins and though pre-eminent in his field never became Registrar-General. He first ventured into the field of occupational-mortality statistics in the Registrar-General's 14th report, for 1851, published in 1855.

became available with civil registration in 1837, Gregory King's impressions were confirmed. For most of Victoria's reign, lung diseases caused a quarter of all deaths, falling to just over a fifth in 1901. Lung cancer, if the diagnoses can be relied on, was then a rare disease, killing little more than 200 persons a year – one per cent of the deaths attributed to bronchitis, or to pulmonary tuberculosis. It now kills 35,000 a year, as many as pneumonia and bronchitis put together, and pulmonary tuberculosis has become an extremely rare disease. Whether the certified cause of death can be relied on is somewhat doubtful. Just as there is a statistical association between lung cancer and smoking, so there has been some correlation in the past with atmospheric pollution. The symptoms of lung cancer are not unlike those of other slow-working respiratory disorders such as bronchitis and tuberculosis, and it is more than likely that doctors unfamiliar with what was then apparently a rare disease, sometimes certified death as due to bronchitis or tuberculosis where a modern doctor would suspect cancer. In so far as there might have been some wrong diagnosis, bronchitis was perhaps not so typically the English disease as the statistics suggest. It remained for all that, a prominent cause of death, carrying off 37,000 in 1951, three times as many as lung cancer, and causing more than six per cent of all deaths. By 1987 fewer than 10,000 died of bronchitis in England and Wales. Bronchitis, a chronic disease, caused its victims much ill health over a long period, and accounted for a high proportion of visits to a doctor. In the years 1951–4 in ten representative general practices, bronchitis, influenza and the common cold came top of the list of reasons for which patients consulted a doctor. Later surveys in 1971 and 1981–2, still showed respiratory disorders as the most common reason for consulting a doctor, though in 1981 they accounted for only 16 per cent of all consultations, a substantial drop from the 27 per cent of ten years earlier.

The causes of a disease like bronchitis are rarely simple. Correlations with smoking, with heavy work, with low social class and with density of population have all been demonstrated. Cold bedrooms have also come under suspicion: it was easier to justify the mania for fresh air when few houses had central heating, and it is unlikely that cold bedrooms could have been explained as a symptom of poverty alone. Of these possible causes, density of population most clearly implies atmospheric pollution, and as recently as 1973 the correlation between density of population and death from bronchitis was very strong. The risk of death from bronchitis increased steadily with crowding: least in rural districts, the risk grew through urban districts with populations of 50,000 or less, through larger urban districts, to the conurbations where death from bronchitis was 40 per cent more likely than in rural districts. While nobody can be sure, it has been estimated that one-fifth of deaths from bronchitis took their origin in atmospheric pollution; nor can there be much doubt that pulmonary tuberculosis, lung cancer and asthma were all aggravated by the insults that dirty air inflicted upon the lungs. More generally, the loss of winter sunlight, still severe in London, Manchester and Glasgow at the end of the Second World War, had a debilitating effect on public health, and the pallor of the people of Manchester in the mid–1950s still contrasted sharply with the healthy complexions observable in a smaller town like Exeter.[15] The rapid elimination of smoke from the skies of Britain within the past forty years is as satisfactory as it is surprising.

REFERENCES

1. A Parker, 'Coal in relation to atmospheric pollution' (Chadwick Lecture, 1945), repr in DSIR, *Atmospheric pollution 1939–45* (1949), p 116; J H Brazell, *London weather* (1968), pp 12, 102, 155; Ministry of Health, *Mortality and morbidity during the London fog of 1952* (1954), pp 33–5, 60–1; TJ Chandler, *Climate of London* (1965), p 213; PRO, MH 58/398.
2. NSCA, 'Sulphur dioxide: summary of a report by the technical committee' (typescript 1971, British Library of Political and Economic Science [BLPES]); DSIR, *Atmospheric pollution 1944–54* (1955), p 15; NSAS, *Proceedings of the annual conference Glasgow 1953,* pp 25–7.
3. Smoke abatement committee, *Official report* (1882), pp 2, 144, 180–1; E Ashby and M Anderson, *The politics of clean air* (1981), pp 83–5, 88, 97; r.c. on coal supplies, *Evidence* (PP 1904 XXII ii) qq 13243 ff.
4. S.c. on the nuisance from smoke (PP 1843 VII), evidence of T M Vickery; Census of Production 1907, *Final report* (PP 1912–13 CIX), p 834; Gas Council, *First report and accounts* (PP 1950–1 XIII), p 46; NSAS, *Proceedings of the annual conference 1932,* p 25.
5. *18th annual report on alkali works* (PP 1883 XVIII), pp 18–19; [Beaver] Committee on air pollution (Interim) *Report* (PP 1953–4 VIII) pp 6, 11, 14; s.c. on the nuisance from smoke (1843), qu 680; A Redford and I S Russell, *The history of local government in Manchester* (1940) III, p 236; J Richards, 'Success with smoke control: Manchester' (NSCA, *Clean air conference 1965*), p 19; A Moss, 'Smoke control – a stocktaking of the present position' (ibid. *1973*) p 3; E W Foskett, 'What has been done – and what remains to be done' (ibid. *1977*) p 2; City of London

(various powers) Act, 1954; *Parliamentary debates*, 3 November, 1955, col 1121 ff.

6. A R Smith, 'The cost of domestic smoke control' (NSCA, *Clean air conference 1967*), p 120; E Burke, 'Dust control in the cement industry' (ibid. *1969*), pp 87 and 92; *73rd report on alkali works*, (1937), p 19, *105th* (1969), p 36.
7. *111th [and last] report on alkali etc. works* (1975), p 13; P A Matthews, 'Iron and steel – progress towards clean air' (NSCA, *Clean air conference 1961*), pp 99–100; E J Challis, 'Clean air – industry's standpoint' (ibid. *1970*) p 63.
8. Gas Council, *First report* (1951); NSAS, *Proceedings of the annual conference 1953*, p 67; *Domestic fuel supplies and the clean air policy* (PP 1963–4 XXVI), pp 9, 15, 19–20; Roy Brooks Ltd, *Mud, straw and insults* (1988), pp 3–14; *AAS*, v.d.
9. A Parker, 'Progress of smoke control' (NSCA, *Clean air conference 1963*), p 14; A R Smith, 'The cost of domestic smoke control' (ibid. *1967*), p 120; A Moss, 'Smoke control – a stocktaking'(ibid. *1973*), pp 3–4, 11–12 ; DSIR, *National survey of air pollution 1961–71*, I, pp 14–18, 128, 137, IV, pp 148–56, II, pp 92–129; Greater London Council [GLC], *Thirty years on: a review of air pollution in London* (1983), p 19.
10. NSCA, 'Sulphur dioxide' (typescript 1971, BLPES); GLC, *Thirty years on*, p 27; *National survey*, I, pp 18–26; Central Electricity Generating Board [CEGB], *Report and accounts,1975–6*, p 24, *1986–7*, pp 6–7; *The Independent*, 6 August 1987, 2 October 1989.
11. Water resources committee, *Third interim report* (Tidal Power) (PP 1920 XXV), pp 3–5; Department of Energy, *Sizewell public enquiry; report by Sir Frank Layfield*, vol 5, ch 59, pp 1–5; CEGB, *Report and accounts 1975–6*, pp 16–20, *1986–7*, p 29.
12. D Hunter, *The diseases of occupations*, (1978), pp 251–3, 258; *Social Trends 1989* (HMSO), Table 9.2.
13. *111th report on alkali works* (1975), pp 35–7; *Industrial air pollution 1975* (HMSO, 1977).
14. D V Glass and E Grebenik, 'World population 1800–1950' in *Cambridge Economic History of Europe* (Cambridge, 1965), ed M M Postan and H J Habakkuk, VI, p 72; *AAS 1989*, Table 2.23; supplement to the *35th annual report of the Registrar-General*, pp lvii, 450–5 (PP 1875 XVIII), to the *65th annual report of the Registrar-General*, part II, chart facing p xxviii (PP 1905 XVIII); Board of Trade, *British and foreign trade and industrial conditions* (PP 1905 LXXXIV), pp 8–20.
15. D V Glass, 'Two papers on Gregory King' in *Population in history* (1965) eds D V Glass and D E C Eversley, p 164; A M Carr-Saunders, D Caradog Jones and C A Moser, *Social conditions in England and Wales*, (1958), pp 235–40; Royal College of General Practitioners *et al.*, *Morbidity survey 1981–2* (HMSO, 1986), pp 21 and 23; *Registrar-General's statistical review for 1973*, part I (B), pp 117–25; A G Ogilvie and D J Newell, *Chronic bronchitis in Newcastle upon Tyne* (1957); C R Lowe, 'Clean air – the health balance sheet' (NSCA, *Clean air conference 1970*), pp 81–2; P J Lawther, 'The effects of air pollution on health' (ibid. *1977*), pp 1–6; G M Howe, 'People, pollution and retribution' (ibid. *1976*), p 10; DSIR, *Investigation of atmospheric pollution 1939–44* (1949), pp 61–2.

The Stennard Works,
Wakefield August 11th 1868

Dedicated without permission
to the Local Board of Health Wakefield
This Memorandum written with Water
taken from the point of junction this
day between the River Calder and the
towns Sewer – Could the odour only
accompany this sheet also it would add
much to the interest of this Memorandum

Ditto Ditto with water taken from
the Mill goit at the same time

C. W. Clay

CHAPTER FOUR

Water Resources under Strain

THE WATER POLLUTERS

Pollution of rivers is nothing new. In the twelfth century the inhabitants of Tavistock threw their rubbish into the Tavy, which luckily for them ran swiftly and did not silt up. The more numerous inhabitants of London were just as careless and from an early date severely polluted the Fleet brook, which entered the Thames where Blackfriar's Bridge now stands. In 1307 the Fleet was no longer navigable.

> By the filth of tanners and such others [it] was more decayed, also by the raising of wharves, but specially by a diversion of the water made by them of the new Temple for their mills standing without Barnard's Castle.

John Stow writing at the end of the sixteenth century reported that the Fleet Brook had had its last thorough scouring in 1502; a further scouring in 1598 failed to improve its condition 'so that the Brook by means of continual encroachments upon the banks getting over the water, and casting of soilage into the stream, is now become worse cloyed and choken than ever it was before'. (Kingsford, *Survey of London by John Stow,* pp 12–13.) In 1388 Parliament was sufficiently concerned at river pollution to impose

---

Opposite:
A letter written in Calder water, 1868. At a time when river water could be used as ink, it might be supposed that counter-measures would soon be taken. However, the cleaning of Britain's industrial rivers is proving a slow business.

the stiff penalty of £20 on those who cast into ditches and rivers near cities, boroughs and towns, 'dung and filth of the garbage and entrails as well of beasts killed as of other corruptions . . . that the air there is greatly corrupt and infect'. (12 Ric II cxiii.) The state of the Fleet Brook in Tudor times shows well enough that law-making and law enforcement were different things. If it was impossible to keep the Fleet brook clean in the sixteenth century, what hope was there for the Thames, Mersey, Clyde and Trent three hundred years later? The growth of population and industry in the nineteenth century put these and lesser rivers in great jeopardy at a time when the prejudice against government intervention was strongest. Nobody objected more vociferously to river pollution than anglers and fishermen threatened by the loss of their hobby, or their livelihood. Yet parliamentary committees set up in 1824 and 1836 to enquire into salmon fisheries largely confined themselves to technical matters – fishing with nets, the close season, the Saturday slap (no fishing from sunset on Saturday to sunrise on Mondays), and poaching. A distinguished witness, Sir Humphrey Davy, admitted that there were hardly any salmon in the Thames and that they were becoming scarce in the Severn, Trent and Avon, but he looked for remedies in fishing practice – the abolition of stake fishing (at the coast with fixed nets), limitations on the size of nets, and the close season. The 1824 parliamentary committee recognised the hopeless nature of the battle between salmon and trade:

> In those rivers on which large commercial cities are situated and on which the interests of manufacturers have led to the expenditure of vast capital, it is not to be looked for that the salmon fishery should flourish; and while it may be from those causes nearly extinct, it would be chimerical to expect that it should ever be restored. Such cases must be obvious, and the committee by no means wish to make recommendations respecting them, which could only end in failure. (S.c. on the salmon fisheries of the UK, 1824, p 145.)

Somewhat inconsistently, the committee, without proposing legislation, advised against admitting poisonous manufacturing waste into salmon rivers. A later enquiry cautiously repeated this warning and mildly suggested that gasworks and some other (unspecified) manufacturers could mend their ways without serious difficulty.[1] Deference to the power of industry and trade could hardly have gone further.

In the prevailing climate of opinion nothing less than a threat to public health could have checked the degradation of British rivers. That public health left much to be desired was undeniable. The first of four cholera epidemics broke out in 1831, but its severity is unknown because it came before civil registration of births, marriages and deaths had begun. The next two outbreaks in 1848 and 1853–4 are much better documented, statistically and by medical and public enquiry. The last serious outbreak of cholera in Britain occurred in 1866 and was largely confined to East London. A further cholera epidemic in continental Europe in 1891 scarcely affected Britain, where urban water supplies were by then satisfactory in quality, and for the most part in quantity as well. Cholera was one symptom, an alarming one, of the poor and uneven state of public health. Civil registration of births, marriages and deaths provided evidence of high death rates in towns, with Liverpool and Glasgow unhealthiest of all. It also became clear that poverty, overcrowding and filth had some as yet unexplained connection with disease and death. The connection between polluted water supplies and cholera was vividly demonstrated in the epidemic of 1853–54 when an unwitting experiment was carried out on the people of South London. Customers of the Southwark and Vauxhall Water Company, drawing its supply from the Thames at Battersea, died of cholera in large numbers. Customers of the Lambeth Water Company were luckier: supplies came from a cleaner stretch of the Thames above Teddington Lock.

Parliament had already demanded that the London water companies should give up the use of the heavily polluted tidal stretches of the Thames near London but not all the companies had complied when cholera returned in 1853. It had never been doubted that clean water was better than dirty; the events of 1853–54 simply convinced the sceptics that cholera must be a water-borne disease. But the germ theory had to wait until Pasteur and Koch demonstrated it in the 1870s. These scientific discoveries did not lead to any improvement in the lower reaches of Britain's rivers because adequate supplies of clean drinking water were available elsewhere. The upper reaches of large rivers like the Thames, the resources of central Wales and of the Peak and Lake Districts, together with artesian wells, dams and reservoirs and the installation of filters guaranteed water supplies of reasonably good quality to the great towns of industrial England. The towns themselves might stand on the black and even stinking lower reaches of the same rivers, but an unsightly stream devoid of fish is not necessarily a threat to health. Events were to show that aesthetic and ecological arguments unsupported by public health considerations or a powerful body of opinion would not persuade local authorities or industry to refrain from river pollution. Indeed concern for public health, far from encouraging the cleansing of rivers, hastened their pollution: as main drainage spread, so more and more urine and night-soil, plentifully diluted with water, poured into the rivers, there to be further diluted. Dilution is often the simplest treatment for waste: high chimneys, water closets, and cooling towers all have a common purpose – the dissipation of harmful or offensive byproducts through dilution.

Industry had long favoured river- or canal-bank sites. Water power, water for steam-raising, and cheap transport were needs common to most firms. Many also needed access to large quantities of water for their manufacturing processes: as a raw material in brewing and distilling; for washing coal, wool and the rags used in paper-making; as a coolant for steam engines and their successors, the steam turbine. After use the water was returned to the river heated, carrying solids in suspension, or with acids, alkali, or other substances in solution. Water pumped from mines often had different characteristics from river water: it could be very hard, and it sometimes carried traces of poisonous metals like arsenic and lead. This water too went into the rivers.

The woollen and worsted industry had no doubt been lowering the quality of river water for hundreds of years but it was not until well into the nineteenth century that serious pollution occurred. The Calder, which ran through the southern part of the West Riding's textile district, remained friendly to fish at least until 1852. A Wakefield shopkeeper who was a keen angler kept a diary of his catches. In 1852, his last good year, he caught 80 lb of chub and dace in a dozen expeditions; in 1853 a similar number of expeditions earned him only 48 lb, but by 1855 he could catch only 14 lb. In 1856 and 1857 he had no luck at all, as he told the royal commission on rivers pollution in 1866. Other witnesses also dated the deterioration of the river to the late 1840s and the early 1850s. Most of the material dredged from the Aire and Calder Navigation was spent dyewood and engine ashes. In one of those telling illustrations that enliven some Victorian blue books, the royal commission reproduced in facsimile a letter written in Calder water sampled at a particularly unfavourable point – where the river received the contents of Wakefield's sewers. The water was strong enough to make a tolerable greyish ink. The author, an agricultural-implement maker, only regretted that he could not send a specimen of the odour of the ink as well. The pollution of the Aire and its tributaries perhaps began a little earlier, but not much. Old men forget and anglers are boastful, but when a considerable landowner and former MP recalled that the Aire was a limpid trout stream as late as 1826 and that serious pollution of its lower reaches began in the 1830s, he is entitled to be believed – he was after all under 60 when he appeared before the royal commission.* Higher up the Aire at Bingley, the river remained reasonably clear of pollution until the 1840s. A Leeds sanitary inspector gave similar evidence: as a boy he caught many a basketful of small fish in the Aire at Leeds up to 1827; ten years later the river had become very foul. Bradford stood not on the Aire but on a tributary, the Bradford Beck. Few streams were its equal for filth in the 1860s, except perhaps the Bradford Canal which, being fed from the beck, had a poor start in life.

> Although it has usually been considered an impossible feat to set the Thames on fire it was found practicable to set the Bradford Canal on fire, as this at times formed part of the amusement of boys in the neighbourhood. They struck a match placed on the end of a stick, reached over and set the canal on fire, the flame rising six feet and running along the water for many yards, like a will-o-the-wisp; canal boats have been so enveloped in flame as to frighten persons on board. (R.c. on rivers pollution, *Third report*, 1867, p xxxviii.)

'With very few exceptions,' the royal commission sadly concluded in 1867, 'the streams of the West Riding of Yorkshire run with a liquid that has more the appearance of ink than of water.'[2] (Ibid. p xxi.)

Much the same could have been said of any river that ran through populous or industrial districts. The rivers of Lancashire were among the worst. The temperature of the Irwell at a weir below Manchester was 20°F higher than the ambient air early one July morning in 1869, and the air was filled with the stench of a 'gaseous emanation' many yards away. The description comes not from the pen of a disgruntled angler, but from Sir Edward Frankland FRS, royal commissioner and distinguished chemist. Salmon no longer visited the Mersey. The Croal, tributary of the Irwell, was polluted by the sewage of Bolton. In Scotland famous salmon rivers were threatened by the discharges of jute and cotton mills. In the Midlands, the Attorney-General secured an injunction in 1858 against the Corporation of Birmingham, which was pouring

* William Ferrand of St Ives, near Bingley.

raw sewage into local streams. The most notorious example of mid-century river pollution was the state of the Thames during a heatwave in 1858. The work of draining the Metropolis was proceeding apace, but the sewage was still being discharged into the river at London. The Houses of Parliament, complaining bitterly of the stench, adjourned for a week, while the Metropolitan Board of Works prayed for rain and poured lime on troubled waters. Trunk sewers carrying London's waste, still untreated, into the Thames downstream at Barking and Crossness, were completed in 1864, and for the next twenty years Barking suffered for the general good.[3] It would be easy if tedious to extend this catalogue of pollution, but more profitable to consider what remedies were proposed and tried at the time.

REMEDIES FOR WATER POLLUTION

Cooperation

Voluntary associations of manufacturers pledged to considerate treatment of rivers were not out of the question. One of the most clearly damaging ways to pollute a river was to throw ashes into it; more cautious firms left the ashes on the river bank knowing that winter floods would carry them away. The practice was not widespread in the 1860s, if the answers to questionnaires distributed by the royal commission can be believed. But it has to be admitted that the response rate was only moderate. In England and Wales only 12 per cent of over 6000 questionnaires were returned; in Scotland nearly 16 per cent out of 2,500. The difference is not large, but suggests that Scottish industrialists were either more conscientious, or more deferential to authority.* Whether non-responders were worse polluters than responders is unknowable; those who did respond included a number of firms who admitted allowing ashes into rivers. Crossleys, the Halifax carpet-makers, had earlier taken the lead in organising a voluntary agreement to keep ashes out of the Calder and Hebble; at one

* A Bristol tanner, declining to give specific information, explained: 'We object to sending anything like a false statement, and we very much doubt if any tanner in the locality will send you a true one, if the details are given'. (R.c. on rivers pollution, *Third report*, 1873, p 230.)

time, the firm admitted, it had put most of its rubbish into the river. Repentance brought few rewards. The voluntary agreement on the Calder ended when a judge held that 20 years' use conferred a prescriptive right to pollute. Firms on the Hebble maintained their agreement, but abandoned fines on offenders. Adverse publicity may have deterred some firms; fear of flooding if the river bed was raised too far was a more powerful argument; and there was always the chance of a small profit from the sale of ashes to brick-makers, or railway-builders. The evidence suggests that the practice of tipping ashes into rivers was in decline by the 1860s but that it was far from over.

Voluntary agreements to end pollution were unlikely to contribute much to solving the problem, given the fiercely independent frame of mind of mid-Victorian business. Firms undoubtedly suffered inconvenience and loss from polluted water. Ashes, mud and even tar from gasworks got into boilers, and textiles were occasionally damaged. Distillers needed to be particularly watchful: Haigs, one of the largest, offered to purify the effluent it poured into the Leven, if it could be assured of pure water for its own use. Most firms resigned themselves to using polluted water, knowing that they would be put to no expense to purify their own effluent. The questionnaires circulated by the royal commission invited proposals for cleansing the rivers, but few firms bothered to make any. Two who evidently had no great fear of competition at home or abroad, were more positive. James Napier, a manufacturing chemist in Glasgow, proposed to ban sewage and manufacturing refuse from rivers altogether:

> If the prohibition were made general and imperative, manufacturers would find other ways to prevent it, the cost of doing so being put upon the manufactured article and the public (the parties benefited eventually) would have to pay. (R.c. on rivers pollution, *Fourth report*, 1872, p 153.)

A Wakefield spinner and dyer, asked if he would object to laws preventing the pollution of rivers, replied more bluntly: 'Not at all; if all we manufacturers were put on the same footing, we should get it out of the public in the end'. (*Third report*, 1867, qu 1491.)

McNaughton and Thom, calico printers of Birkacre, near Chorley, put forward more comprehensive proposals. If manufacturers were given three years' notice of a ban on pollution, they could find ways to comply: 'What would now be impossible, practically speaking, would be easy if not profitable, in a few years'. (*First report*, 1870, p 172.) They proposed a system of inspection by central government, the inspectors to be men of some standing; and compulsory purchase of land needed for filtration or other works. To safeguard the manufacturing interest they would allow 'best practicable means' as a defence in law, and they proposed that new works could be opened only with the consent of the chief inspector. Some of these ideas are reminiscent in their technological optimism of the anti-smog legislation of modern California. But it is clear that few firms shared this faith in technology or were willing to be put to much expense for the common good.[4]

Litigation

If sewage authorities and manufacturers would not mend their ways voluntarily, could aggrieved parties look to the courts for a remedy? In principle the law of nuisance could bear down on injury or discomfort arising from pollution. Public nuisance, where the injury was general, was a criminal offence punishable by fine; private nuisance gave rise to a civil action, with the twin remedies of damages and abatement. Both kinds of nuisance were known to the common law by the thirteenth century. In the nineteenth century statutory nuisances were created under the public health and similar acts, with the result that sanitary or alkali inspectors and other public servants took action where private persons would previously have been put to the trouble and expense of litigation or prosecution. Over the centuries, notable legal victories were won against polluters. In 1628, in *Jones* v. *Powell*, Chief Justice Hide declared:

> The erecting of a common or private brewhouse in itself is not a nuisance, nor the burning of sea-coal therein; but if it is erected close to the dwelling of another, as here, so that thereby his goods are spoilt and his house rendered uninhabitable, an action will lie.

In 1834 in *R* v. *Medley*, the Equitable Gas Company was indicted for nuisance. Owing to mechanical failure, the company's gasworks:

> injuriously conveyed by divers pipes etc., great quantities of filthy, noxious, unwholesome and deleterious liquids, matter, scum and refuse resulting from the washing of the said gas . . . into the said river Thames.

Thousands of fish died in a stretch of river between Hungerford Market and Millbank. Although fish in the Thames were already in decline, the company and its officers were found guilty and fined.

The case of *Crossley* v. *Lightowler* 1867 established the important point that even a small contribution to a large nuisance might be actionable. Lightowler erected new dyeworks on the Hebble in 1864 at the site of an old dyeworks abandoned 25 years earlier. Crossleys, the Halifax carpet-makers sued, arguing that Lightowler could not avail himself of the defence of a prescriptive right to pollute the Hebble. Lightowler responded that he was by no means the only polluter, but his argument did not satisfy the court and he was restrained by injunction from his offensive practices. In *Young* v. *the Bankier Distillery Co* 1893, the House of Lords came down firmly on the side of clean, unaltered water. Young, a colliery owner, pumped hard water from his mine into a soft-water stream, altering its character. The Lords admitted a prescriptive right to pollute – though it did not apply in this case – but in the words of Lord Macnaghten:

> A riparian proprietor is entitled to have the water of the stream, on the banks of which his property lies, flow down as it has been accustomed to flow down to his property, subject to the ordinary use of the flowing water by upper proprietors . . . without sensible alteration in its character and quality.

In *Rylands* v. *Fletcher* 1868, water from Ryland's reservoir escaped down old workings and flooded a working mine belonging to Fletcher. The House of Lords upheld the decision in the Court of Exchequer Chamber and quoted with approval the ringing words of Mr Justice Blackburn:

> The person whose grass or corn is eaten down by the escaping cattle of his neighbour, or whose mine is flooded by the water from his neighbour's reservoir, or whose cellar is invaded by the filth of his neighbour's privy, or whose habitation is made unhealthy by the fumes of his neighbour's alkali works, is damnified without any fault of his own; and it seems but reasonable and just that the neighbour who has brought something on his own property, (which was not naturally there) harmless to others so long as it is confined to his own property, but which he knows will be mischievous if it gets on his neighbour's, should be obliged to make good the damage which ensues if he does not succeed in confining it to his own property.

A lawyer from Mars, studying these judgments, might well have supposed that pollution would soon become a thing of the past, even when all allowances were made for a prescriptive right to pollute. A glance at the smoky air of London or at the inky waters of the Irwell would have taught him otherwise. A barrister complained to the committee on noxious vapours in 1862 that Islington's varnish makers had been declared a nuisance twenty years before (II Carrington and Payne 485) but were still flourishing. It took the Duke of Buccleuch a generation to get a judgment against the paper-makers on the north Esk near Edinburgh. And the royal commission, while acknowledging the theoretical strength of the law against pollution, concluded that the law was totally ineffective:

> A person judging from the present appearance of the streams in the West Riding . . . would conclude that there existed a general licence to commit every kind of river abuse. (R. c. on rivers pollution, *Third report*, 1867, pp li–lii.)[5]

Royal commissions

Concern about river pollution led to the appointment of a royal commission in 1865. Government had not yet formed the habit of appointing a royal commission when it wished to avoid taking awkward decisions; twenty years later perhaps, fifty years later almost certainly, government would have chosen commissioners representing every interest and shade of opinion, and virtually forced to write a balanced and lukewarm report in the interests of unanimity. The sturdy early Victorian tradition of appointing men who would report with vigour and in forthright terms remained powerful in the 1860s. The commissioners were experts: Robert Rawlinson was an ebullient civil engineer with long experience of water and drainage works, J T Way an agricultural chemist, and J T Harrison another civil engineer. In 1868 the commission was reconstituted with new members: Sir William Denison of the Royal Engineers, a retired colonial governor (as chairman), Edward Frankland, chemist and FRS in 1853 at the age of 28, and J C Morton, an agricultural writer and teacher, and one-time secretary of the Royal Agricultural Society. Denison died in 1871 and was not replaced. Both commissions took extensive oral evidence, issued questionnaires, conducted experiments and published uncomfortable reports. The bent of their minds is well illustrated by their reaction to the views of a powerful committee of millowners. The millowners, who claimed to represent firms with a gross output of £117 million a year, rejected all schemes to limit industrial pollution of rivers:

> Any material change would involve such outlay and destruction of existing property as to be practically confiscating it on measures, which might after all fail to secure the purification of the water after use.

Industrial pollution, they asserted, was not dangerous to health and should not be prohibited by legislation, but left to the arbitrament of successive users. They did support the diversion of sewage from rivers to farms by means of irrigation. The royal commission, with more courage than political sense, brushed aside these objections:

> We cannot understand that the magnitude of a nuisance is any reason for continuing it; neither can we admit that any man can plead the enormous extent of his business as giving him a right to injure his neighbours.

They proposed drastic remedies for the evils of river pollution. Land irrigation with sewage and trade waste was a much canvassed alternative to pouring matter into rivers, and it was no accident that an agricultural expert sat on both commissions. Little industrial waste was toxic at that time, and the waste water from tanneries and the pot ale from distilleries were even richer in nitrogen than town sewage. It therefore seemed feasible to recommend land irrigation and a ban on the admission of either sewage or trade waste into rivers. This was Rawlinson's radical and probably impractical recommendation in his first report in 1867. Frankland's recommendations a few years later have a more modern ring. He proposed standards for the quality of discharge water – with specific limits on the quantities of mineral or organic matter either in suspension or solution, limits on acids, alkalis and arsenic, and a general ban on the discharge of any liquid 'which shall exhibit by daylight a distinct colour'. The deceased chairman Sir William Denison had favoured local enforcement of the law; his successors, in this respect if in no other, agreed with industrial interests in recommending inspection by central government.[6]

Legislation

The clever men from London had done their best;* it remained to be seen how far government and Parliament would proceed along the road that they had charted. Bills failed to become law in 1872 and 1873 and it was left to Disraeli's government to pilot a bill through Parliament in 1876. It is doubtful whether the bill would have done much damage to industrial interests if it had passed through its various stages unscathed. For not only did it allow 'best practicable means' as a defence to the discharge of noxious liquids, it also gave the Local Government Board the last word on prosecution; 'which Board in giving or withholding their consent shall have regard to the industrial interests involved in the case and to the circumstances and requirements of the locality'. In this version much would have depended on the way in which the Board interpreted its responsibilities. The manufacturing interests were not ready to leave such matters to a government department. First, the textile industries obtained an amendment which required the Board to refuse its consent to prosecution in a textile district unless satisfied that the anti-pollution measures 'are reasonably practicable . . . under all the circumstances of the case, and that no material injury will be inflicted by such proceedings on the interests of such manufacture'. In the end all manufacturing districts were given the protection of this wrecking amendment, which admittedly referred only to liquids and to particles in suspension. Industrial effluent was of course one of the main river pollutants. Solids were banned from streams – cinders and the solid refuse of manufactories and quarries were specifically mentioned. Sewage was also to be kept out until purified by best practicable means; however, the Board was empowered to allow sanitary authorities more time for compliance. Whether Parliament intended 'time' to mean 'more than a hundred years' is somewhat doubtful. Enforcement of the law was left to those same sanitary authorities who happened to be among the principal offenders against clean river water. The anomaly persisted until the recent creation of the National Rivers Authority in 1989. Perhaps it is fitting that Charles Darwin's countrymen should move at this petty pace: if Nature does not advance by leaps and bounds, why should government when trying to protect the

* John Ingham, a Yorkshire dyer, told the commission in 1866: 'I always think we are open to receive information. I certainly do not think we are as clever as they are in London, where they have no manufacturers.' (R.c. on rivers pollution, *Third report*, 1873, qu 9244.)

environment? The act provided for the appointment of an inspector to advise the Local Government Board on technical matters such as the best practicable means to purify sewage, or the effluent from a paper mill. Angus Smith, consulting chemist and part-time chief alkali inspector, added this to his other duties.[7]

The Rivers Pollution Prevention Act, 1876, remained the principal statute until 1951. The President of the Local Government Board in presenting the measure to Parliament modestly described it as a skeleton bill, a first attempt at legislation on a difficult subject. Within a few years it was widely condemned as a failure. Perhaps the verdict was a little unfair, for who could say how bad the rivers might have become if Parliament had passed no law at all? The major local authorities in the West Riding of Yorkshire showed some concern to enforce the act, for in 1894 the county council and the county boroughs joined forces and obtained an act of parliament that established the West Riding of Yorkshire Rivers Board. The Board's members were appointed from the ranks of county and county-borough councillors and the Board was given the sole power to prosecute cases of river pollution in its area. Its 11 inspectors had over 200 sewage works, up to 800 mills and many collieries to inspect. Coal-washing had recently become a common practice and until it was regulated great quantities of fine coal dust held in suspension poured into rivers. By a mixture of prosecution and persuasion, the Board slowly got colliery managers to mend their ways and by 1900 all collieries with coal-washing plant had installed settling tanks. If the coal dust recovered from the settling tanks had been less valuable it would have been that much harder to effect the change. Manufacturers who saw no profit in anti-pollution measures such as filters and settling tanks showed little willingness to co-operate with the Rivers Board. A similar body undertook the supervision (conserving would be too flattering a term) of the Mersey-Irwell river system. The Thames Conservancy, a sleepy body, had presided over the degradation of the Thames since 1857. In 1908 control of the tidal reaches of the Thames, below Teddington Lock, passed to the Port of London Authority. But the PLA was more interested in docks than in pollution and a further 75 years passed before an angler could boast that he had caught a salmon with rod and line in the Thames.[8]

THE STATE OF THE RIVERS

In rivers where salmon could live, there is no reliable evidence that fish stocks were dwindling at the end of the nineteenth century. The Fishmongers Company collected statistics of salmon delivered to Billingsgate, but they by no means recorded the whole story, omitting salmon sent direct to private fishmongers, to large provincial towns, and abroad. The figures, for what they are worth, are more notable for violent short-term fluctuations than for any long-term trend. As many boxes of salmon were delivered in 1884 as in 1834, and as few in 1899 as in 1839 – the high figures being almost exactly double the low ones. Salmon prices at Billingsgate were static from 1894 to 1897, but had increased by a quarter (to 1s 4½d. per lb!) in 1899, a year of high prosperity as well as of low fish deliveries. Prices had been much higher in 1825 (another prosperous year), but in 1834 when the catch was large, prices were at about the same level as in 1899 when the catch was low. The story is complicated by imports of Norwegian and Canadian salmon, which were substantial in the late 1890s. These additional supplies, though less highly prized than Scotch salmon, tended to keep prices down; otherwise, a probably static supply of salmon from Scotland would have fetched higher prices as population and average incomes rose in the markets served by Billingsgate. Supplies of Scotch salmon on average did not vary much from the levels of the 1890s for the next twenty years; in the 1920s the catch was rather higher. The salmon is so fastidious about his breeding waters that the evidence shows conclusively that the upper reaches of Scottish salmon rivers remained limpid and free of pollution; what more can be deduced about the state of the rivers of Scotland is unclear. The number of distilleries on the Spey doubled between 1850 and 1900 without any adverse effects on the salmon, and as we have seen, the murky waters of the lower Thames did not deter a few intrepid salmon from making their way upstream in 1983 to the clearer waters beyond London, Reading and Oxford.[9]

TREATMENT OF SEWAGE

The sorry state of the major rivers of the British Isles at the end of the nineteenth century led to the appointment of a further royal commission. This royal commission on sewage disposal began its

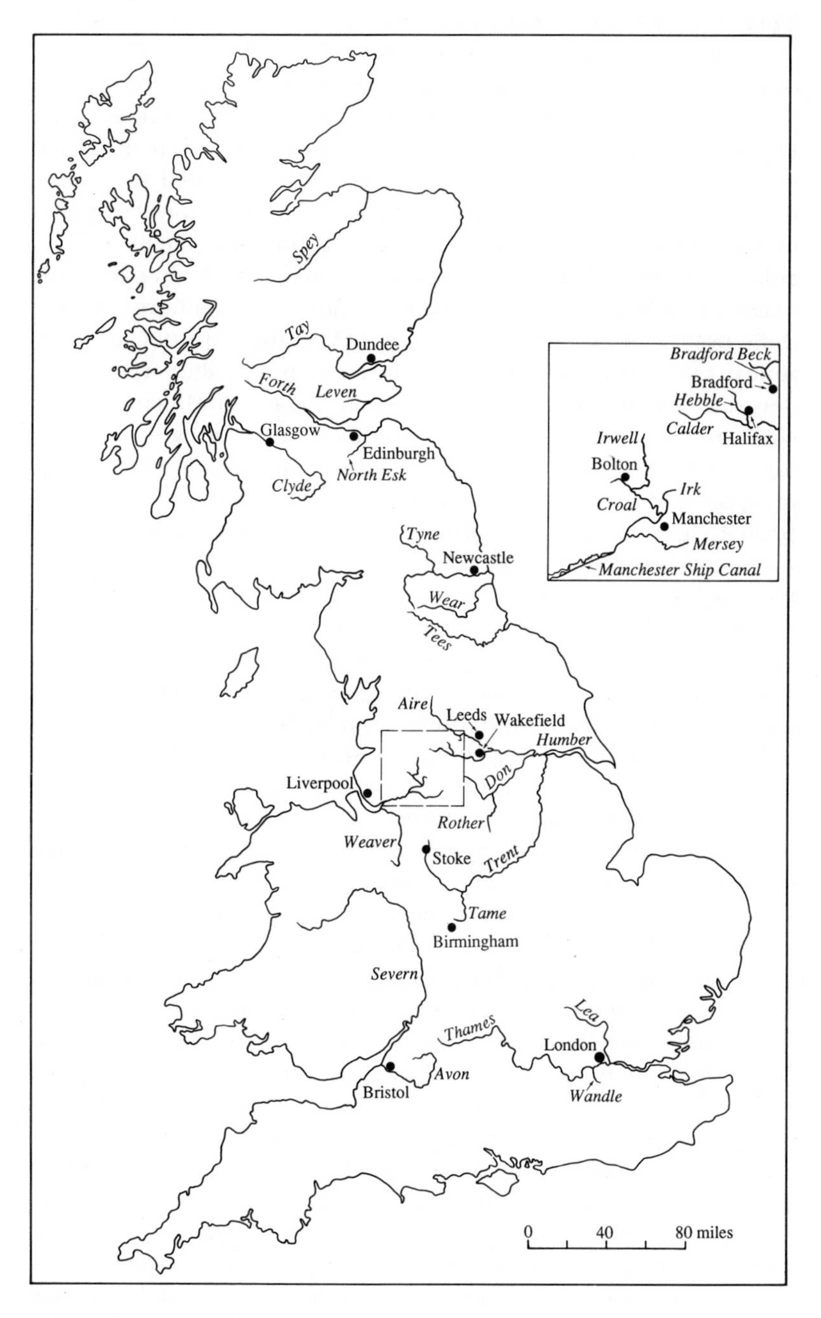

Map 2: The polluted rivers of Britain

labours in 1898 and issued its tenth and final report 17 years later in 1915. Like the earlier commissions in the 1860s and 1870s, it was composed of experts and like them it enquired not just into sewage disposal but into industrial pollutants as well. Unlike Rawlinson and Frankland however, the new experts were political realists. For example, when confronted with shellfish contaminated by sewage, they recommended not measures to abate the pollution, but power to river boards to control the taking of fish and shellfish! They tailored their recommendation to local conditions – river flow, the needs of industry, and the finances of the local authority. Unlike their predecessors in the 1860s and 1870s, they did not aspire to return rivers to their original condition of purity. Rather, they laid down moderate standards to which treated effluent should conform, in the hope that as time went by, the standards could be tightened up. They recognised that under the terms of the 1876 act manufacturers had no incentive to undertake research into preventing pollution, but that sewage authorities were perhaps somewhat more amenable to the pressure of opinion. Wherever there was main drainage, that is to say, wherever there were towns, rivers were at risk, and there can have been few that were not more or less polluted in this way. At the time that the royal commission was sitting, new methods of treating sewage were coming into use and the enquiry, by proposing standards, encouraged sewage authorities to strive for better results, and gave the Local Government Board a measure of performance to which it might coax sewage authorities to conform.

At the end of the nineteenth century, a good many methods were in use for the disposal of sewage. The simplest was to pour it untreated into a large body of water. Towns on the coast or on tidal stretches of river were usually tempted to follow this course, Liverpool, Hull, Bristol, Torquay, Yarmouth and Lowestoft among them. Lowestoft's sewage was discharged into the sea 'at a point which does not affect the bathing' – as the Lowestoft entry in the Municipal Year Book was careful to point out. More than a quarter of the authorities giving details to the Year Book used this method for at least part of their sewage. Landlocked or more conscientious authorities had a harder task. Put simply, their problem was to separate the solids from the liquids, and to make both comparatively innocuous before discharge – of the liquids to the river or sea, the solids to land or by dumping at sea. The easiest part, though a noisome one, was separation of the solids: this was achieved by the use of coarse filters and settlement tanks, assisted by chemical precipitation. Lime and salts of iron were the preferred precipitants. The sludge deposit was pressed to rid it of excess water. Ambitious authorities went further and dried it, hoping to sell it as fertiliser. Others were content to bury it on land, or ship it to sea and dump it at a reasonable distance from the shore. The liquid part of sewage though heavily diluted with water from water closets and surface-water drains needed a great deal of treatment before it could be safely poured into rivers.

The Rawlinson and Frankland commissions, following best contemporary opinion, favoured the establishment of sewage farms. Under this system the local authority would buy land, preferably to the north-east of the town and at least a mile away, and establish a farm irrigated by sewage, crude or separated. In the end the sewage would find its way riverwards, filtered by the earth and oxidised by the air to a sweet-smelling harmless liquid. Bacteria would have played a part in the process of purification, but that was better understood in the 1880s than in the 1860s. Croydon, then a small town separate from London, was running a successful sewage farm in the 1860s and the fashion rapidly caught on in the next generation. In 1875, 87 towns were treating their sewage in this way. More joined in later. Not all the farms were genuine agricultural holdings. The system of 'intermittent filtration' took liquid sewage only; it needed less land than a sewage farm proper; but as a result of more intensive use, the land was rarely cultivated. Where crops were grown, under what was called 'broad irrigation', they were lush and heavy, without being profitable. They often left a surplus of receipts over current expenditure, but when the capital costs of pumping sewage were allowed for, a loss resulted. Sewage farms were, nevertheless, a cheap way of dealing with the problem and it is not surprising that they attracted an enthusiastic following, especially among the smaller urban authorities set in ample countryside. Some large towns – Manchester and Nottingham, for example – also had extensive farms at the end of the nineteenth century.

The sewage farm could be regarded as a natural way of filtering, aerating and purifying effluent. The same effect could be achieved artificially. Already in 1870 Leamington was experimenting with animal charcoal or burnt bones as a filter for sewage. Contact beds filled to a depth of three feet or more with slate or coke were a development of this system. By 1900 acknowledged to be the best method of aerating and purifying sewage, they were still much less popular than the sewage farm or the use of chemical precipitants, but they held out better prospects of a reasonably satisfactory effluent. Later refinements such as the use of compressed air (the activated sludge process) promoted more thorough oxidisation and so reduced the demand on the oxygen dissolved in river water. Trade waste from dyeworks, distilleries and tanneries was often a more powerful pollutant (took more oxygen from water) than sewage. Where local authorities admitted such trade waste into their sewers, and had the necessary contact-beds for filtration, they were able to limit the damage to river quality. Under the Public Health Act 1875 local authorities were empowered, but not obliged, to accept trade waste, and to levy an appropriate charge. In 1937 they were compelled, under certain safeguards, to admit trade waste, so enacting recommendations first made in 1903 by the royal commission on sewage disposal.[10]

Comparison of the enquiries of 1865–74 and of 1898–1915 gives the impression that as in a well-played game of chess, every move to abate river pollution was countered by another that increased the strain on the system. Technically much good work had been done, especially to limit the harm caused by sewage. Many firms too had adopted some process for purifying waste, but it was often relatively ineffective. In any case, the scale of the problems had grown with the increase of industrial production and with the spread of main drainage and the water-closet system. In both world wars the condition of rivers deteriorated: prevention of pollution had a low priority in industry and little money was available for the upkeep of sewers and sewage works. Per capita expenditure on sewage works in real terms had still not recovered to the prewar level in 1953, though it advanced strongly in the later 1950s and 1960s. Unfortunately it is not possible to disentangle money spent on sewers from money spent on sewage disposal – only the latter can affect river quality. From the collapse of the boom in 1973-4, capital spending steadily shrank until by 1981–2 it was only half what it had been eight years earlier. There has been little improvement since – which explains why the recently privatised water companies are faced with a heavy bill for capital works.

THE LAST SIXTY YEARS

These adverse circumstances go far in explaining why rivers remained seriously polluted and deadly to fish until very recently. No comprehensive survey was undertaken for more than 40 years after the commission on sewage disposal published its final report in 1915, but the condition of the Trent and its tributaries may fairly represent the industrial rivers of Britain between the wars. The Trent Fishery Board found that in 1936, out of 550 miles of river, nearly a quarter (130 miles) was lethal to all animal and plant life. Another 50 miles permitted the lower forms of plant life to exist, but not fish or insects. The Trent, which rises a few miles north of Stoke, had the misfortune of being heavily polluted on its southward course almost from its birth. For thirty miles in and below the Potteries, it was a dead river in 1936. It had just begun to recover as it turned to travel north when it received the waters of the Tame just after they had passed through Birmingham and the Black Country. After this setback, recovery was bound to be slow. The river next ran the gauntlet of Derby and Nottingham. Meandering north through Lincolnshire it became tidal at Gainsborough. But that was not the end of its troubles for in joining the Yorkshire Ouse (to become the Humber) it was also receiving the inky waters of the Aire and Calder. Two modern pollutants – cooling water from power stations and nitrates from farmland – were not a serious threat in 1936. To a considerable extent, however, electricity replaces steam power in industry. The condensing water used to cool steam-engine cylinders heated river water in exactly the same way as power stations sited on big rivers and estuaries do today, but on a smaller scale. It has been estimated that at the beginning of the twentieth century, factory steam-engines consumed some 50 million tons of coal, little more than half the coal and oil required by the boilers of modern power stations.[11]

Much information about the state of rivers was collected in 1958 in the first of a series of official surveys. The results need careful handling. Judgement of river quality is not an exact science, though objective tests are included in the process, and coverage includes some surprisingly insignificant streams – those with a summer flow of one million gallons a day (mgd), or little more than two inches of water in a channel two feet wide travelling at 4 mph! The published data make no attempt to allow for the relative size of rivers, with the result that one mile of the Thames immediately above Teddington Lock counts equally with a mile of the Exe at Dulverton.* Whether the absence of weighting exaggerates or understates the extent of pollution is unclear. On the one hand, the sources of rivers are usually to be found in thinly peopled upland

* Half the length of rivers surveyed had a summer flow of less than 3 mgd.

districts where pollution used to be rare. It is on the middle and lower reaches of non-tidal rivers that towns and industry cluster and pour out their waste products, and it is these stretches that contain large volumes of water. On the other hand, with tidal rivers it is usual for great cities like London and Glasgow to be sited a considerable distance from the sea, with the result that the heaviest pollution occurs, or used to occur, in the smaller body of water. As the river widens on its approach to the sea, the pollutants are much diluted with sea water and river quality is likely to improve. Yet a mile of the Thames near London Bridge counts in the statistics for as much as a mile of the Thames at Canvey Island. The authors of the survey recognise this problem of weighting but offer no solutions.

Further complications arise from the changing pattern of pollution. Before 1945, British farms raised fewer cattle and pigs than they do today, and less animal manure had to be disposed of; nor were so many tons of artificial fertilisers, especially nitrates, applied to the land. The pollution of water by the escape of slurry or the leaching of nitrates hardly occurred; in recent years such incidents have become common and agriculture is now an important cause of water pollution. Since agriculture is practised almost everywhere, no stream is now entirely free from the risk of pollution. A contrary tendency has been at work to divert town and industrial waste away from the river on the doorstep to the lower reaches and to the estuaries. This practice began long ago when the crude sewage of London was diverted to Barking in Essex and Crossness in Kent, some ten miles downstream from London Bridge. Glasgow and Manchester later followed the same policy, the sewage of Manchester passing into the ship canal and that of Glasgow into the lower Clyde. Since the Second World War the steel and chemical industries, oil refining and electricity generation have all been attracted to estuarine or coastal sites. In 1970 for example, two-thirds of industry's cooling water (mostly but not entirely from electricity generation) was discharged into tidal, but necessarily estuarine, waters; a further sixth was discharged into the open sea.

The discharge of other industrial effluent and of sewage and sewage effluent followed a different pattern in 1970. The dry weather flow of effluent (dwf) at 4,000 mgd was little more than a quarter of the volume of cooling water discharged (15,000 mgd). Yet it still represented a formidable mass of polluted water – nearly three times the average flow of the Thames over Teddington Weir.

Relatively little industrial effluent passed into the sewers to be processed in sewage works; much (about 2,000 mgd) was returned direct to rivers and canals. Unlike cooling water, over three-fifths (64 per cent) of industrial and sewage effluent passed in 1970 into non-tidal rivers and just over a third (35 per cent) into tidal waters. About a half of the industrial effluent failed to reach a satisfactory standard, that is, it did not comply with the standards set by the river authority. Compliance by sewage authorities was even rarer: only 40 per cent of sewage effluent was satisfactory in the eyes of the river boards. A majority of their members were appointed from the ranks of the local authorities, which must go far in explaining why unsatisfactory discharges were tolerated on a scale that would have made an alkali inspector blush. The sewage of four-fifths of the population of England and Wales was subject to these standards, which for most discharges were those recommended by the royal commission early in the century. For some discharges the standard was higher than the royal commission's. For a smaller number the standard was lower; even here, over half the discharges were unsatisfactory. The remaining fifth of the population, some ten million people, discharged their untreated crude sewage into tidal rivers (four million) or into estuaries or the sea (six million).*

The river quality surveys of 1958 and 1970 show that tidal rivers, despite the larger volume of water they contained, were more likely to be polluted than the non-tidal stretches. By 1990, when the latest survey was undertaken, the tidal rivers had apparently become cleaner than the non-tidal. This sharp improvement in the quality of tidal rivers owes something to investment in new treatment plants; it may also owe something to a change in the way river quality is classified – and to that extent the improvement is merely statistical and does not represent any physical change. Extrapolation from the trend between 1958 and 1970 would certainly not give the 1990 results for tidal rivers.

* The dumping of sewage sludge into the North Sea is to end by 1998, but some sewage effluent will continue to be discharged into the sea after the year 2000.

Table 4.1 River quality in England and Wales 1958–90 (percentage of length)

	Good	*Doubtful*	*Poor*	*Grossly polluted*
Non-tidal				
1958	72	15	7	6
1970	74	17	5	4
1980	69	21	8	2
1990	63	25	9	2
Tidal				
1958	41	32	14	13
1970	48	23	17	12
1980	68	23	5	4
1990	66	24	7	3

Grossly polluted stretches of non-tidal river (by definition lethal to fish) have diminished, in accord with what is known from other sources, while stretches of doubtful and poor quality river water have tended to increase – a trend compatible with the overloading of sewage works and with the effects of intensive agriculture. But even if it is admitted that the tidal rivers are less clean than the 1980 and 1990 figures suggest, there can be little doubt that by mileage and probably by volume too, clean rivers are commoner than dirty, and are indeed the norm. Pollution is largely confined to the industrial rivers of South Wales, the Midlands and the North. There has been substantial improvement in the condition of the Thames. In 1970, the non-tidal stretches of the Thames and its tributaries were already reasonably clean, but the tidal Thames from Teddington Lock to the sea was in a sorry state. Hardly any of the river was grossly polluted, but four-fifths was poor (that is, had substantial biochemical oxygen demand and contained toxic pollutants and solids in suspension); the rest was doubtful – had reduced oxygen content and was subject to toxic discharges. By 1990, the tidal Thames was judged to be in good condition, except for the long stretch from the Tower of London to a point due south of Basildon; this stretch, about half the length of the tidal Thames, was adjudged 'fair', the term that has replaced 'doubtful' in the classification of river quality. The Thames at Battersea, the cause of cholera in 1853–4, is now judged to be of good quality; it still looks murky and uninviting, but the banks are clean when visible at low tide except for pieces of plastic, and the river allows passage to some salmon. A third of the country's water supply is drawn from such sources as the lower stretches of rivers, that is from rivers containing treated sewage effluent. There is no risk to health in drinking such water – nor is there much pleasure, and it is hardly surprising that the demand for bottled water has soared in a fastidious and affluent population.*

Improvement has been slower for the other industrial rivers of England and Wales. The Tame, dead in 1936 and for many years before, had benefited from the closure of some old sewage works. In 1990, parts of it supported, rather uncertainly, some coarse fishing. The Trent below its junction with the Tame was in reasonably good condition and well stocked with fish. Some parts of the system remained poor or bad, and Severn-Trent, along with Yorkshire, North-West (and South-West!) were the only regions in 1990 where more than ten per cent of the rivers were of poor or bad quality. Rivers that had attracted much adverse comment in the 1860s and 1870s – the Mersey, Irwell, Weaver, Don, Rother, Aire and Calder – remain in poor condition for much of their length. Except in tidal waters, little improvement is discernible in the past thirty years, and only a great deal of money and pressure will restore these rivers to good quality in a reasonably short time. In the 1960s, when industry was spending large sums to abate atmospheric pollution, little capital was invested in clean water. According to a CBI survey, spending by industry in the decade came to £10 million, and the operating costs of pollution equipment amounted to £20 million a year. It is hardly surprising therefore that industrial rivers continued to be seriously polluted. Since then the law has changed, but pollution goes on.

If laws made rivers clean, the period 1948–74 would have marked the end of river pollution in Britain. To judge from the statute book, a long period of legislative and administrative neglect was coming to an end. The 1876 act gave little encouragement to authorities who might wish to seek out and prosecute offenders, and only the West Riding River Board had any stomach for the fight. Between 1930 and 1947, the Ministry of Health received 44 applications to prosecute, two-thirds of them from the West Riding Board. Cases were often settled out of court if the offender agreed to undertake remedial work. The Ministry hardly ever refused to allow a prosecution in the other cases. After the war a series of acts

* The use of ozone instead of chlorine in water purification would improve the flavour of tap water and the bank balance of water drinkers. Mineral water is still a minority taste: in 1989, the average Englishman drank only five pints of it; in 1991, sixteen pints or less than a glass a week.

– in 1948, 1951, 1961, 1963 and 1973 – betokened an increasing interest in water purity and resources. River catchment boards, where they had been established since 1930, became river boards in 1948, and where no catchment boards existed, new river boards came into being. In 1963, the boards were renamed river authorities and in 1974 amalgamations took place that resulted in the creation of ten water authorities for England and Wales, privatised in 1989. Some of these changes amounted to little more than a change of name – for example the replacement of catchment boards by river boards. Others were more substantial. In 1951 the river boards took over responsibility for enforcement of the pollution laws from local authorities, under the new Rivers (Prevention of Pollution) Act. The act replaced the 1876 act and gave river boards somewhat greater powers than the sanitary authorities had possessed. For example, the boards had the power to prescribe standards for effluent, as recommended long before by the royal commission on sewage disposal. But time for compliance was allowed, and the standards had to have regard for the character and flow of the stream and for the interests – industrial, agricultural, or fishing – that used it. Where there were no standards, pollution was forbidden unless the polluter had taken 'all reasonably practicable steps' to prevent it. New discharges of effluent needed the consent of the river board. Only in 1963 were pre-1951 discharges brought within the consent procedure. The 1951 act which had repealed a principal statute 75 years old, itself lasted only 23 years before being replaced in turn by Part II of the Control of Pollution Act, 1974.

This seemingly majestic statute had little substance until expanded by ministerial administrative orders. So far as river pollution was concerned, few orders had come into force within the next few years. In any case, fundamental weaknesses in the administrative structure continued as they had ever since 1876; the sewage authorities, heavily represented on catchment and river boards and, after 1974, on the new water/sewage authorities, were prime polluters of the rivers they were empowered and obliged to cleanse. The creation of a National Rivers Authority in 1989 has at last separated law enforcement from river management. It remains disquieting and odd that two conflicting functions, water supply and sewage disposal – are carried out by one and the same body. Despite appearances, it would be wrong to describe the law as nothing but a handsome facade concealing filthy slums. Not only has there been undeniable if slow improvement in the state of some rivers; there has also been much greater willingness to prosecute offenders. Whereas there was barely one prosecution a year between 1930 and 1947, there were over 300 in 1987, mostly of farmers and industrialists. Sewage authorities, responsible for over 4,000 cases of pollution, were prosecuted only seven times in the year. In all, more than 20,000 incidents of pollution were recorded in 1987, half attributable to industry and agriculture, a fifth to sewage works. It cannot be said that a regime of one prosecution for every sixty offences is unduly severe, only that it is severe by previous standards, reflecting rising hostility to environmental degradation. In 1989 David (the villagers of Wargrave in Berkshire) successfully prosecuted Goliath (the Thames Water Authority), and the National Rivers Authority promised rigorous enforcement of the law. The newly privatised water companies are being allowed to raise their charges by much more than the rate of inflation, in order to enable them to invest in new plant for the treatment of sewage and industrial effluent. With the increasingly active European Economic Community also taking an interest in Britain's environment, it is no longer unreasonable or wildly optimistic to hope that our rivers, estuaries and coastal waters may yet return in due course to something like their original purity.[12]

REFERENCES

1. W Camden, *Britannia*, ed E Gibson (1772 edn), p 158; C L Kingsford, *Survey of London by John Stow* (1908), pp 12–13; 12 Ric II c XIII; s.c. on the salmon fisheries of the UK (PP 1824 VII), App. 3, p 145; ibid., *Second report* (PP 1825 V), p 4; s.c. on the salmon fisheries, Scotland (PP 1836 XVIII), *Report*, pp vi–vii.
2. R.c. on rivers pollution, *Third report: evidence* (PP 1867 XXXIII), qq 1067–86, 1249, 2918, 5483–7, 6578, 11824, 11925, 11952–4; *report*, p 95; [2nd] r.c. on rivers pollution, *Third report* (PP 1871 XXV), p 12; *Third report* (PP 1867 XXXIII), pp xxi, xxxviii.
3. [2nd] r.c. on rivers pollution, *First report* (PP 1870 XL), pp 16 n, 18, 138, 147; *Fourth report* (PP 1872 XXXIV), pp 94–5, 100, 119, 148, 154; R A Buckley, *The law of nuisance* (1981), p 123; W A Robson, *The government and misgovernment of London* (1939), pp 126–7.
4. [2nd] r.c. on rivers pollution, *Evidence* (PP 1873 XXXVI), pp 80–242; *Fourth report* pp 43, 72, 153; [1st] r.c. (PP 1867 XXXIII), qq 1490–1, 11014, 11086, 11097, 11107, 11157, 12008 ff, 12114–9; [2nd] r.c. (PP 1870 XL), p 172.
5. E W and H G Garrett, *Law of nuisances* (1908), pp 1–2, 6; *English reports*, vol 172, p 1247; *Law Journal Reports, Chancery* vol 36, 1866–7, pp

584–90; *Law Times Reports,* 24 February 1894, pp 838–41, and 7 July 1868, pp 220–3; Lords committee on noxious vapours (PP 1862 XIV), qq 2033 ff, 2062; r.c. on rivers pollution, *Third report* (PP 1867 XXXIII), pp li–lii.

6. [2nd] r.c. on rivers pollution, *First report* (PP 1870 XL), pp 42–3; *Evidence* (ibid.), pp 289–90; *Third report* (PP 1871 XXV), p 56.
7. The bill is in PP 1876 V; *Parliamentary debates,* 11 July 1876, col 1282, 20 July cols 1675–7, 24 July cols 1875–9, 25 July col 1882, 4 August cols 556–61; 39 and 40 Vic c 75.
8. R.c. on salmon fisheries, *Evidence* (PP 1902 XIII), qq 20884–20978; A Redford and I S Russell, *The history of local government in Manchester* (1940), II, pp 393, 400, III, pp 113–16, 118; *General report of the conservators of the River Thames 1900* (PP 1901 LX); *Municipal yearbook for 1985,* p 584; 57 and 58 Vic ch clxvi (a local act).
9. R.c. on salmon fisheries *Report,* (PP 1902 XIII), pp 20–23, 44, and Apps VIII and IX; Fishery Board for Scotland, 46th report (PP 1928 IX), pp 87–8.
10. R.c. on sewage disposal, *Fourth report* (PP 1904 XXXVII), p xxi; *Tenth report* (PP 1914–16 XXXV) summarises earlier recommendations; *Municipal yearbook for 1898;* W Robertson and C Porter, *Sanitary law and practice* (1912), pp 433–48; H Stephens, *The book of the farm* (1855), II, 441–2; r.c. on rivers pollution, *First report* (PP 1866 XXXIII), pp 12, 15; *Third report: evidence* (PP 1867 XXXIII), qq 14658, 15000; [2nd] r.c., *First report* (PP 1870 XL), pp 41–51; s.c. on municipal trading, *Evidence* (PP 1900 VII), qq 3216–20; r.c. on salmon fisheries (PP 1902 XIII), qq 21669–21763; Public Health (Drainage of Trade Premises) Act, 1937; *Parliamentary debates,* 27 May 1937, col 540 ff; r.c. on metropolitan sewage discharge, *Second report* (PP 1884–5 XXXI), p xxxviii.
11. Ministry of Health, *Prevention of rivers pollution* (1949), p 16; PRO, HLG 50/2714; Department of the Environment [DoE], *Rivers pollution survey 1970* (1971), I, p 3 ff; DoE, *River quality in England and Wales 1985* (1986), pp 11 and 16; *The Listener,* 22 July 1936; r.c. on coal supplies, *Final report* (PP 1905 XVI), p 12.
12. DoE, *River pollution survey 1970* (1972), II, pp 1–5, 43; ibid. (1974), III, p 3; ibid., (1971), I, p 3,12–3, 24–5; DoE, *River quality in England and Wales 1985* (1986); National Rivers Authority, *The quality of rivers, canals and estuaries in England and Wales* (1991); *The Independent,* 4 March 1989; *Social Trends 1989,* Table 9.19; Ministry of Health, *Prevention of rivers pollution* (1949), p 34.

The beginning of the M40 at Uxbridge, west of London. Less well-known than Spaghetti Junction near Birmingham, the beginning of the M40 illustrates just as forcefully the large demands that motorways make on our most precious natural resource, land.

CHAPTER FIVE

Land Loss and Reclamation

Air is free, and the owner of a river bank has only tenuous rights over the water that flows past his property. What nobody owns may be polluted with impunity, unless the law forbids it and is rigorously enforced. Luckily air and water are in constant movement and shed their pollutants easily: they fall from the sky, are deposited as silt, or are swept out to sea. There is therefore a natural tendency for air and water to cleanse themselves if given a chance. Ill treatment of land has far worse consequences: in the short run, land is fixed both in amount and position; a municipal rubbish tip, a slag heap, or twelve feet of alkali waste will ruin it for years. A road, a factory, or a housing estate will also ruin it for agricultural purposes, but by adding value to the land will not arouse quite the same sense of loss. In the long ages of geological time wind, rain and frost can obliterate any damage, but the process is immeasurably slower than the cleansing of air or water. Land has however one important protection that is lacking to air or water – it is not a free good, but for the most part (in Britain) is owned either privately or by a corporation. To some small extent land can be injured by atmospheric pollution or by toxic flooding, but on the whole it is fair to say that if a firm, or public authority, intends to change the character of land it will first have to buy it; in other words, it will incur costs and not evade them by leaving the community to pick up the bill, as happens with the pollution of air or water. The study of land loss and reclamation will therefore take a different course from that followed in preceding chapters. The emphasis will fall not on pollution but on changing land use.

COASTAL EROSION

Loss of land to the sea has been rare in historic times, and is largely confined to the east coast. Dunwich in Suffolk is probably the best known casualty. A medieval town of some importance, it has since mostly disappeared beneath the waves. At the time of the Reform Bill of 1832 it was notorious as a rotten borough with a few resident electors and the economy of a small village. Other vulnerable stretches of the east coast in Lincolnshire and the East Riding also suffered long-term loss. The Holderness coast in the East Riding receded on average in the century after 1850, by about 130 yards and as a result the sea has swallowed about 1,800 acres of land. If the rate of erosion had been constant since the time of Julius Caesar (a large assumption), about 55 square miles or 36,000 acres would have been lost. The Holderness coast from Flamborough Head to just north of Spurn Head at the mouth of the Humber measures about 35 miles. If the whole coastline of Britain receded at that rate the results would be serious, but as it happens losses round Britain are comparatively rare. The problem is slight in Scotland and for most of the west coast of England and Wales. There is some erosion on the south coast, for example at Lyme Regis and Charmouth where the cliffs are of soft clay, and further east at Brighton. In the very long run the problems of southern and eastern England will grow if the downward tilt of the land continues. So far as the movement can be measured the south-east appears to be sinking at the rate of a foot a century. Freak weather conditions pose a more serious problem. In January 1953 severe northerly gales caused a surge in the North Sea and sea defences were breached at many places on the east coast. Three hundred people were drowned and over 200,000 acres of land flooded. Had the surge coincided with high tide and with rivers in full spate, much greater loss of life and damage would have occurred. London escaped flooding on that occasion and has since built the Thames Barrage, which should protect it for the next 30 years – at the expense of smaller communities and less valuable property downstream, that stand to be flooded if the flow of the tide is impeded.

Erosion is not necessarily all loss. Some of the material carried away from Holderness will have sunk to the sea bed, but it is probable that a large quantity has been carried ashore at Spurn Head or further south in Lincolnshire to strengthen the natural defences against the sea there. Left to itself the sea will usually give back part if not all of the land it has eroded. Where man interferes with this natural process by building sea walls and groynes he prevents erosion at (say) Lowestoft or Brighton but in so doing causes it further along the coast. Lowestoft and Brighton have kept the beach sand that would otherwise have been carried by the sea to places further south or west. Those places also suffer loss but now without compensation, for the sea defences check the natural transfer of sand and pebbles. It would be too costly, and probably futile, to offer protection to all the vulnerable stretches of coastline. Apart from safeguarding the major towns, there is little to be done except to forbid the removal of sand or pebbles by contractors, for their interests plainly conflict with attempts to conserve the shore. The Coast Protection Act of 1939 empowered local authorities to forbid the extraction of beach material except under licence; until then no general power existed to control its removal.* Indeed, the inhabitants of Devon and Cornwall had a general licence to take sand for the improvement of tillage, though not for other purposes. (The licence dated from a grant made by Richard, Earl of Cornwall, in 1261.) It is unlikely that human depredations can compete with the power of the sea in causing land loss. Neither agency, according to the best estimates that can be made, has altered the coastline much in recent times. A royal commission in 1911 assessed losses in England and Wales in the previous sixty years at no more than 5,000 acres and a further 2,000 acres in Scotland and Ireland together. 'The danger', they declared, 'is by no means alarming.' In any case the gains, for what they were worth, comfortably exceeded the losses. The most substantial additions were in Lancashire and Lincolnshire, with smaller gains in Norfolk and Yorkshire.

LAND RECLAMATION

That gains should exceed losses was only to be expected for they comprised not only material cast up by the sea but sand and silt brought down by rivers from hill and dale. In Morecambe Bay, for example, large areas of saltmarsh tempted projectors to put forward schemes of reclamation. But the sand brought down from the hills of the Lake District proved a poor basis for agriculture and much of

* The Harbours Act, 1814, gave the Admiralty power to prohibit the removal of beach material in the vicinity of harbours.

the land reclaimed was later abandoned. Norfolk offered a more promising field of endeavour. In 1837 a company was mooted with the purpose of reclaiming 150,000 acres in the Wash at a cost of two million pounds. A less ambitious project for reclaiming a mere 32,000 acres was accepted by Parliament in 1846. The Norfolk Estuary Company undertook to cut a four-mile drainage channel from King's Lynn to the sea in exchange for parliamentary approval. The channel cost £280,000 and took eleven years to complete. Instead of reclaiming 32,000 acres in six years the company had reclaimed only 2,200 acres in sixty. It had begun life with plenty of capital and little experience; sixty years later it had experience, but no money. Costs of reclamation at £25 an acre equalled the value of the land when it was fit for sale after a long period. The original projectors had grossly underestimated the time needed to rid the land of salt, and found themselves waiting many years for any return on their capital. Had they been engaged in forestry the initial investment would have been much lower and there would have been some return from thinnings at an earlier stage. At the best of times forestry requires patience, faith and an acceptance of low profits; land reclamation in the Wash proved even less rewarding.

Inland drainage, though not without its problems, was more profitable. The 1,300 square miles of fenland in East Anglia were more thoroughly drained in the nineteenth century than ever before, thanks to the steam engine, which continued to pump water away after shrinkage of the peat had upset the natural drainage levels. The extensive wetlands of the Somerset Levels were reclaimed for highly productive agriculture at the same time. Not that the unimproved wetlands had been without some agricultural value. They had provided summer grazing, reeds and wild fowl; thorough drainage increased agricultural productivity; it did not create land or value where none had existed before. The draining of the Fens is therefore unlike the recent Dutch schemes for reclaiming the Zuyder Zee, which when complete will have farms in what had once been a shallow sea.[1] It is, rather, a spectacular example of the improved drainage of agricultural land that was widely practised in Victorian Britain, allowing cold wet lands to bear earlier and bigger crops.

EROSION INLAND

Traditionally, north-west Europe has been free from the risks of land erosion that occur where temperatures are very high, or where exceptionally heavy rainfall is common. In Africa and North America, where climatic conditions are less favourable, erosion can be prevented only if agriculture and forestry are conducted with great prudence: over-grazing, the ploughing of slopes, and monoculture, all need to be avoided. Severe erosion may be disastrous for one farmer or a district, but it may not represent a total loss to agriculture. If a Kansas farm loses topsoil in a dust storm the dust presumably comes to rest somewhere else in America; if heavy rain washes the soil into rivers it will come ashore downstream or at worst at the mouth of the Mississippi, where in the fullness of time it will become dry and fertile land once more. This is no comfort to the Kansas farmer and not very much to farmers anywhere, so that in practice nobody favours erosion though many take too few precautions against it.

In Britain it would be hard to find references to erosion in agricultural writings before the middle of this century; the worst, it was believed, that could happen to soil was that it would become infertile with poor agricultural practice. A rotation of crops including roots and clover, a plentiful supply of farmyard manure, together with artificial fertilisers and drainage where necessary, raised agricultural productivity substantially in the nineteenth and early twentieth centuries. Mechanisation and the heavy use of artificials between the wars and especially since 1939 have changed the face of agriculture and brought the possibility of erosion somewhat closer. The need for a rotation of crops was an article of faith with theorists and most farmers until after the Second World War. The faith was sometimes violated between 1939 and 1945 in the desperate need to maximise output during the war, and more persistently after the war when large doses of artificial fertilisers appeared to make monoculture feasible. The peaty soils of the Fenlands, rich in humus, lent themselves best to the new system of farming. Eastern England, though the home of the Norfolk rotation with its alternation of corn, roots and clover, had moved away from temporary grasses even before the war. Over the country as a whole, 31 per cent of the arable was under temporary grasses in 1936; in eastern England it was less than half that percentage. Since the war, eastern England has had less and less land under temporary grass: the prewar pattern still held immediately after the war, but by the mid-sixties the proportion of temporary grasses had fallen to less than ten per cent; by 1987 Norfolk had the highest percentage of temporary grasses – a mere 3.7 per cent; Lincoln, Suffolk and Cambridge had less – Cambridge only 1.5 per cent. A visitor could be forgiven for thinking that the park surrounding Holkham Hall was the only large area of grass in a region given over to cornfields. In fact other and equally exhausting crops were grown as well, sugar beet and potatoes. In Britain the number of cattle has increased by a half since the 1930s but in eastern England the number has fallen – not so far as the temporary grasses, it is true, but any fall reduces the supply of farmyard manure. If any soils can afford to do without the organic matter provided by temporary grasses and farmyard manure it is the peaty soils of the Fens, which are by definition largely of organic origin. Even these soils are not inexhaustible and they will slowly wear out and even disappear if subjected to this treatment for too long. The practice of stubble-burning removes another source of organic matter. Stubble-burning has now been banned by law; but another danger will take its place if farmers are tempted to sell their straw for paper-making instead. Eurocell, a subsidiary of British Sugar, is proposing to establish a £300 million plant on Humberside for turning straw into paper. If successful the project will save 300,000 acres of tropical rainforest every year, but it may be contributing in a small way to the destruction of some of the richest agricultural land in England.

Over the country as a whole, temporary grass remained an important part of the farming system, 24 per cent of arable in 1987. But it had been declining for the previous ten years, having held steady at about 30–2 per cent of arable for many years up to the late 1970s. If farmers everywhere think tomorrow what eastern England thinks today, the amount of temporary grass will dwindle ominously in the next generation, bringing nearer the prospect of lower fertility and perhaps erosion. Other modern practices reinforce this possibility. The use of heavy machinery compacts the soil so that it holds less water, and loose soil is more likely to be carried off in heavy rain. Ploughing up and down slopes is safer for tractors and drivers than ploughing along the contours, but more likely to provide channels for water-borne erosion. Hedgerows act as windbreaks and collect wind-blown soil at their base. The removal of hedgerows, which has proceeded apace in post-war Britain, destroys a feature of the landscape that is not only picturesque and a haven for wildlife, but also an important defence against wind erosion.[2] Over large tracts of land where parliamentary enclosure took place 150 or 200 years ago, the hedgerows date only from that time or later. Now the wish or the need for quick profits is leading to their wholesale destruction. It is not necessary to adhere fanatically to the organic-farming movement to see that some tendencies in modern agriculture take little thought for the morrow. It is doubtful if high profits and high output are compatible with treating the land as a part of the national heritage. If farmers are to act as trustees or stewards, and not treat the land as an expendable commodity like a tractor, they may need some metaphorical hedges of their own in order to break the force of the winds of competition.

FORESTRY

Erosion, whether by the sea or inland at the hands of farmers, has so far affected little land, and is a problem for the future rather than the present. One possible defence against coastal erosion, widely canvassed early this century, was forestry. It had long been suspected, though it was hard to prove, that trees had a beneficial effect on the land: they attracted rain, held on to water and offered strong resistance to erosion. In well-wooded countries like France, Germany and the United States the preservation of forests seemed a more important and desirable objective than in countries like Britain where woodland had largely given way to arable and grass. It was an American diplomat, scholar and politician, George Perkins Marsh, whose concern for the vanishing forests of his native Vermont led him to write a lengthy manifesto of conservation, *Man and Nature or Physical Geography as Modified by Human Action.* First published in 1864, in London as well as in New York, *Man and Nature* went through several editions in Marsh's lifetime and was reissued in 1964 on the occasion of its centenary. Widely influential in America the work seems largely to have gone unnoticed in Britain in the author's lifetime. The longest chapter in the book and large parts of other chapters are devoted to the theme of forest destruction, so that for all its learning the work cannot be regarded as a balanced account of man's use and misuse of natural resources. In particular, Marsh pays little attention to mineral resources, which to him were apparently limitless. In Britain, well-endowed with coal rather than timber, W S Jevons's *Coal Question* published a year later caused a much greater stir. Even the Admiralty lost interest in timber in the 1860s, for the iron warship made the supply of Baltic pine and of home-grown oak a subject of indifference to the controllers of naval dockyards.

Since Tudor times the supply of naval timber had been a matter of keen anxiety in England. The iron industry of the Sussex and Kentish Weald with its insatiable demand for charcoal was thought to threaten the survival of timber essential to naval shipbuilders. Country gentlemen like John Evelyn felt they ought to plant trees not just for profit but as a duty to the navy and to posterity:

> Highly necessary it were that men should perpetually be planting, that so posterity might have trees fit for their service, of competent, that is, of middle growth and age . . . I know it is an objection or rather an unreasonable excuse of the slothful neglect of successive and continual planting, that the expectation is tedious of what is not likely to be timber in our time. (Evelyn, *Silva*, pp 217–18.)

Woodland, even before the days of the Forestry Commission, was not all oak, and the naval dockyard was not its only destination. House-building, pitprops, furniture-making, the domestic fire and the industrial furnace were some of the many other uses of wood. And for them beech, elm, ash or pine might be serviceable. Nor was it necessary to go to the trouble of planting. When a tree is cut down its roots do not die and new growth is only a matter of time. Foresters had long taken advantage of this fact by managing their woodlands as coppice. Under this system, regenerated trees naturally, either from stump or seed, form a stand which is felled comparatively young and then allowed to regenerate once more. Some of the young trees may be spared to grow to maturity, a system known as coppice with standards. A well-managed coppice can therefore supply wood indefinitely without the need for clear felling.

Country gentlemen valued trees not only for income but as an amenity. A stand of trees provided cover for foxes and for game birds, though it was difficult to have both at the same time since the fox was an enemy of the pheasant. No park was complete without trees carefully planted to adorn the view, but the fond owner might shrink from felling them at maturity, preferring to let them die, worthless, of old age. The acreage of woodland, itself a doubtful statistic when first collected in the late nineteenth century, is therefore no guide to the extent of forestry conducted on economic principles. It would appear that Britain then had about

2.75 million acres of woodland and plantation. Well over half was situated in England and most of the rest in Scotland. Broad-leaved deciduous species predominated since the woods were for the most part ancient and on the whole represented what naturally grew best, not what gave the best economic return. The most careful of the early censuses of woodland was taken in 1905. Under four per cent of the woodland, it appeared, had been deliberately planted in the previous ten years (104,000 acres); a similar acreage had been planted in the 'eighties and it therefore seems reasonable to suppose that owners replanted less than 0.4 per cent of their woodland every year. If all woodland had been managed under the system of plantation and clear felling, this proportion would have been much too low for profitable forestry – it would have taken over 250 years for complete replanting. Since the slow-growing oak reaches maturity in 120 or 150 years, and beech and ash in about 80 years, stands would have been well past their prime when due for felling. However, the system of coppicing was popular in England, if not in Scotland or Wales, and a third of English woodland was so managed.[3] If coppicing is excluded from the calculations (nearly 600,000 acres in all) nearly five per cent of the remaining woodland was replanted every ten years; even so, it would still have taken over 200 years to replant the whole. From the point of view of economic forestry, woodland management in Britain clearly left much to be desired.

Few people worried about British woodlands, either on economic or strategic grounds. Imports of timber grew rapidly in the later nineteenth century, trebling between 1864 and 1899 when they reached 10 million loads (roughly 10 million tons).* In addition there was a large and growing import of wood pulp for paper-making. It was estimated that self-sufficiency in timber products from temperate climates would require a further nine million acres of woodland, that is a quadrupling of the forested area. The Admiralty was confident that the Royal Navy could keep the seas open and gave evidence to this effect to the royal commission that considered the supply of food in time of war. The Board of Trade and most economists saw no danger in relying upon foreign supplies of food or timber since Britain had a healthy surplus on its balance of payments and could usually afford to make substantial foreign investments and loans out of that surplus. Those who wished to encourage forestry therefore had to rely on social and

* A load of timber is 50 cubic feet; 66 cu ft of sawn softwoods, or 44 cu ft of pitwood, or sawn hardwoods, weigh a ton.

geographical arguments. It was known that trees held land together and were a powerful protection against the movement of sand. The Earl of Leicester planted pines at Holkham for ornament and windbreak and was pleasantly surprised to discover that they checked the movement of sand as well. The Liberal government that came to power in 1906 was anxious to increase opportunities for work on the land. It therefore amended the terms of reference of the royal commission on coastal erosion so that they should include some study of afforestation as a source of employment. As a result the commission's second report was largely devoted to this topic. They proposed a large annual planting of trees – 150,000 acres – but the government responded cautiously. The Development Commission set up in 1909 was given many duties, among them research into forestry and the acquisition of land with a view to planting. Forestry was not high on its list of priorities and little had been done in this area, apart from some encouragement to forestry education, at the outbreak of the First World War. Heavy shipping losses and a desperate shortage of foreign exchange showed that the prewar complacency about timber imports was ill-founded. A national policy for woodland took its place among the many plans for post-war reconstruction, and the Forestry Commission began work in November 1919.

The long-term objective of the new policy was to secure a three-year supply of home-grown timber against any future emergency. To this end the commission was given a substantial grant of money with which it hoped to plant over a million acres in 40 years and 1.75 million acres in 80 years. More immediately, it looked to plant 150,000 acres in ten years and to encourage the replanting and good management of the country's private woodlands, which at 3 million acres were much more extensive than the largest acreage the commission itself ever hoped to plant. Heavy wartime demand had resulted in a great deal of felling of private woodland, and when the commission completed its survey of forestry resources in 1925 it was disappointed to find that the condition of private woodland was poor and deteriorating. Encouragement of private forestry took the form of a grant of £2 per acre towards the cost of planting and replanting. Though modest in amount and hedged round with restrictions, the grant was taken up in respect of 77,000 acres in the first ten years. In that time the forestry commission acquired 300,000 acres of plantable land (mostly heath and moorland) and from 1923 took charge of the Crown's ancient forests amounting to 120,000 acres. In order to obtain the land it needed, the commission often had to buy farms together with land of little or no economic value, though possibly of scenic beauty. In its first ten years it planted a somewhat smaller acreage than planned – 141,000 acres, of which only 7,000 acres were planted with broad-leaved trees. It was coniferous trees that the market wanted, and the commission would have got few thanks from industry or parliament if it had behaved like an eighteenth-century lord, planting oaks for posterity, amenity and the Navy. Where poor heathland was being planted for the first time few objected to a coniferous forest, but where land was of good quality and had previously grown beech or oak (for example on the Haldon Hills west of Exeter) many regretted the intrusion of the dark coniferous forests.

An attack from another quarter on grounds of public economy was repulsed with some ease. The [Geddes] committee on national expenditure worked with all the bustle of which five leading businessmen were capable, and published three reports well within seven months of being appointed in 1921. They recommended the abolition of the forestry commission on the grounds that the strategic arguments for home-grown timber applied equally to wheat, clothing and armaments. In their haste, they forgot that a field of grass could be ploughed up to give a crop of wheat the following year, whereas it took eighty years to grow a mature tree. The government, anxious though it was to placate the opponents of high public spending, had no intention of sacrificing forestry. In the next financial crisis, in 1931, the [May] committee on national expenditure proceeded with more caution, merely recommending some curtailment of the commission's activities; government accepted only some of these proposals and the commission emerged from the crisis virtually unscathed.

The business of afforesting waste proceeded rather faster in the 1930s and in the early years of the war. Planting and replanting were restricted from 1942 to 1946, but were in full swing again in 1947. By then the commission had afforested or reafforested half a million acres, including 38,000 acres of hardwoods. The commission's plantations were too young to make any useful contribution to demand in the Second World War and it was older private woodlands that replaced, as far as possible, imported timber and pulp. A sixth of private woodland, nearly half a million acres, was clear-felled or 'devastated' during and immediately after the Second World War. By 'devastated' was meant the cutting of all valuable timber, leaving behind scattered trees, birch, or coppice.

There has, justifiably, been little confidence in a steady surplus on Britain's balance of payments at any time since the Second World War. Where a surplus is unlikely, import-saving becomes a favoured policy, and for a generation after 1945 forestry received ample encouragement from public funds. As a result the forestry commission more than quadrupled the acreage it had planted before 1947. Until the early 1970s it was planting more than 50,000 acres a year; since then the pace has slackened considerably and has recently been running at something less than the prewar level. In 1960, the commission announced that it was virtually abandoning the planting of hardwoods – a decision that it reversed in 1974 in the face of mounting protests and with some shift of emphasis in policy. Import-saving remained a weighty consideration, but to it was joined a concern for amenity and scenic beauty that led to the planting of rather more hardwoods. At the same time, the operations of the commission, like those of other nationalised industries, were subjected to stricter financial discipline. In principle, the forestry commission had always expected, and been expected, to pay its way in the long run. Unlike other businesses its accounts were closed not once a year but once in eighty years; in the meantime the original costs of acquisition and planting of land were mounting up at compound interest. Foresters took it as a general rule that periodic thinnings would cover the cost of forest maintenance, and that the final clear-felling after eighty years would recover the original costs (swollen though they were by compound interest, traditionally set at 2.5 or 3 per cent). The economics of forestry had been worked out in nineteenth-century Germany at a time of stable prices and low long-term rates of interest. Applied to late twentieth-century Britain with inflation rates ranging from 5 to 25 per cent a year and with long-term rates of interest rarely below ten per cent the sums of money notionally at stake became very large, and the commission's liabilities looked alarming. For example, when the costs of establishing a plantation amount to £1,000, with compound interest at 2.5 per cent, the forester has to raise £7,200 to cover his costs by clear-felling after eighty years. But when interest is reckoned at ten per cent he will need £2 million! In the light of these puffed-up figures it was decided in 1972 to require a real rate of return of three per cent from the commission rather than the going nominal rate. Either system might well yield satisfactory results in the long run. Substitutes are unlikely to be found for timber in all its many uses, and whatever the cost there is likely to be a market for the forester's products. High nominal interest rates reflect high inflation: the high prices that inflate the forester's liabilities will also inflate his receipts when clear-felling occurs. There is, consequently, no good reason to think that the economic results of forestry will in the long run turn out any better or any worse under inflation than under price stability. Those who dislike the economic calculus and are pleased to see two blades of grass grow where one grew before will take heart from the addition of 2.5 million acres to the country's forests in the past 70 years, mostly planted on land for which there was little other use.

The state's forests are the largest in the country, but private woodlands in the aggregate still account for more than three-fifths of the forested area. They take up less space in this brief narrative only because it is always easier to document public than private enterprise. The grants in aid of planting available to private owners between the wars covered about a quarter of the cost of establishing a plantation, and were not attractive enough to encourage much replanting at a time of falling timber prices. In 1944 the Commission devised a scheme for 'dedicated woodlands' under which proprietors undertook to manage their woodland to a high standard in return for generous grants. The scheme became law in 1945 and grants amounting to about half the cost of planting became available. In its early years the scheme attracted little support because landowners distrusted the austere and radical Labour government and because, until late in 1949, controls kept the price of timber unprofitably low. With the ending of price controls and under a more sympathetic government, owners took heart and enthusiastically applied for the grants – for management as well as planting. Now, nearly half of the three million acres of private woodland come within the dedication scheme.[4] Britain is still far from self-sufficient in timber and timber products, and it can be argued that the money spent on reclaiming wasteland for forestry would have been better invested in manufacturing industry or financial services. However, the costs of promoting forestry have never been a great burden on the exchequer and if every penny had been diverted to investment elsewhere (a highly improbable event!) it would have added little to the resources of industry or of the City. The performance of the British economy does not inspire much faith in its ability to forgo the help it has received and will receive from import-saving measures. Timber being one of the raw materials whose cost, in real terms, has risen in the past hundred years, it seems very likely that governments have done well to develop forestry after a long period of neglect.

LOSS OF AGRICULTURAL LAND TO OTHER USES

Domestic refuse

Land drainage and afforestation have added substantially to the agricultural resources of Britain at a time when there have been heavy losses of agricultural land to tipping, mineral excavation, building and roads. It is now time to examine these losses. Tipping of domestic refuse is as old as history and rubbish tips have provided archaeologists with much valuable evidence. The changing contents of Britain's dustbins still provide an interesting commentary on economic life, but it is the weight and volume of refuse that is of concern here. Large-scale tipping in modern times may be said to date from the time of the mid-nineteenth-century sewage farms. The sewage farm was a comparatively short-lived experiment confined to a particular kind of refuse. Other domestic refuse, once the domain of the private contractor, gradually came to be the responsibility of local government under the Nuisance Removal and Public Health Acts. Whereas the private contractor made his living by sorting and selling whatever was useful in his Dickensian mounds, by 'recycling' to use a modern term, the local authority approached refuse from the point of view of public health. Recycling was only a minor consideration, and tipping often proved the most convenient method of disposal. The nuisance that tipping could cause – flies, rats, smells, fire and wind-blown rubbish – was recognised early, and by the 1890s incineration was becoming an increasingly popular alternative. Several engineering firms marketed patent 'dust destructors' – Meldrum, Horsfall, and Manlove and Alliot prominent among them. Borough surveyors were assured that they could get a useful amount of heat – for example, for generating electricity, for disinfection by steam, or for pumping sewage – from otherwise useless refuse. In time the ambitious attempt to make incineration pay was abandoned and the incinerator came to be used as a convenient way of lessening the volume of refuse for tipping or 'landfill', the fashionable and not entirely inaccurate euphemism for tipping.*

---

* Domestic refuse suffers badly from the English habit of calling a spade a teaspoon. The rubbish tip has become a civic amenity site; refuse collection is now solid-waste management; and the rag-and-bone man works in the materials reclamation business. Perhaps the ostrich should replace the lion as a symbol of national character.

The weight and, even more, the volume of refuse increase as time goes by. When most homes had a coal fire it was easy to burn newspapers and wrappers. As coal fires disappeared, so did the ashes and cinders, but the volume of paper and plastic wrapping in the dustbin more than offset the missing cinders. In 1893, 12 per cent of London's refuse was paper and vegetables; in 1979 60 per cent, and a further 23 per cent was plastic, glass and metal. Estimates of the weight of domestic refuse are somewhat unreliable because local authority collections include waste from shops and offices. It would appear that local authorities now handle from 17 to 18 million tons of refuse a year, of which about four-fifths is domestic. This amounts to some 15 hundred weight per household. New Yorkers, even more extravagant, generate a similar quantity per head, destined like most British refuse for landfill. Sixty years ago local authorities were collecting about 10 million tons of refuse, and where to tip it had long been a problem. Boroughs were becoming more fastidious than Plymouth had been in the 1870s, and took good care to export their rubbish to country districts just as their sludge was tipped at sea or otherwise disposed of at a comfortable distance. Willesden in Middlesex was old-fashioned enough in 1914 to let out refuse collection to a private contractor, but at the same time shrewd enough to stipulate that there was to be no tipping within the borough boundaries. In the 1920s the London boroughs had such a poor reputation for careless and inconsiderate tipping that the Ministry of Health instituted an enquiry.

The least offensive place to choose for tipping is a hole in the ground – true landfill – and the least nuisance will arise if the rubbish is compacted and covered with a layer of earth. Marshy land, gravel pits and disused quarries have been favourite sites, and the extensive brickworks of the south Midlands offer a convenient source of landfill for local authorities over a wide area. Since the brickworks plan to continue excavating clay for many years, the local authority is assured of adequate landfill for the foreseeable future. The County of Avon has embarked on one such scheme. It has entered into long-term agreements with British Rail and with Shanks & McEwan, a leading waste contractor. Avon compacts its waste before packing it into containers for carriage by British Rail shuttle service to the Shanks & McEwan landfill site – brick excavations – at Calvert in Buckinghamshire. Every working day a train leaves Bristol, stops to pick up more containers at Bath and proceeds to Calvert. The void space available at present amounts to 15 million cubic metres and in time a further 37 million cubic metres of clay will be excavated, providing space for Avon's rubbish at least until the end of the century. The London borough of Hillingdon also uses the Calvert site. Eventually the land will be restored, probably to agriculture. Such schemes are expensive but when well managed offer a useful future for derelict land without creating too much nuisance in the meantime.

Industrial refuse

Local authority refuse is only a small part of the waste seeking a resting place every year. In modern Britain, huge quantities of material – spoil from mining, slag from steelworks, earth and rock from building operations, and material of all kinds from manufacturing industry as well as domestic refuse – have to be disposed of. Whether the total weight of refuse is rising or falling is hard to determine. The output of coal is now lower than at any time since the 1850s, and only a sixth of what it was at its peak in 1913. Colliery slag heaps cannot be covering the face of the land as rapidly as in the past. Even so it is estimated that 100 million tons of waste arise in a year in the coal industry. Other industries generate in total twice as much solid waste as coal mining. The steel industry has shrunk by about a third since the early 1970s, but still produces twice as much steel as in 1913. The china clay industry of Cornwall has created a ghostly and not unattractive landscape of spoilheaps wherever clay workings are found. For every ton of kaolin produced, about 8 tons of overburden and coarser-grained clay have to be separated out and dumped, no mean quantity now that output of china clay exceeds four million tons a year. Power stations burn most of Britain's coal, and inevitably accumulate a small mountain of ash every year. A small part of these waste materials have a use other than as mere landfill. Blast furnace slag derived from phosphoric ores is ground up to make fertiliser, and some of the waste from clay workings and some power station ash is converted into building panels or cement. The rest has to be dumped: if there is space close at hand as with clay workings the firm will itself undertake the dumping at relatively low cost. If the waste has to be moved a considerable distance, waste contractors are likely to be called in. The word 'skip' has gained a new meaning with the introduction of metal containers for builder's rubble, collected by sturdy lorries for removal to a landfill site. The 1932 edition of the Oxford English Dictionary did not record this meaning, but Chambers had noted it by 1976 and the new OED (1989) also has it. Previous usage confined skips to the mining industry.

Building is one of many industries unlikely to have enough space of its own for the dumping of waste. The problem is not a new one. It was affecting the alkali manufacture more than a century ago. Few of the Widnes firms had ground on which to tip their alkali waste, but one of them – John Hutchinson and Co. – had land at Ditton Marsh, and for 7½d. to 9d. a ton would accept other firms' waste. By the 1890s some 500 acres had been buried 12 feet deep in alkali waste estimated to weigh ten million tons. (Every ton of soda generated over four tons of solid waste under the Leblanc process.) The whole alkali trade produced nearly 300,000 tons of soda in the early 1860s and doubled its output in the next generation. The Ditton Marsh tip was a large one but took only a fraction of the alkali waste of south-west Lancashire. When carefully sealed after tipping it caused little nuisance, though its sulphur content (up to 15 per cent) effectively killed all vegetation for many years. When carelessly dumped, the nuisance from sulphuretted hydrogen became unendurable.[5]

Hazardous waste

In modern parlance, alkali waste, if still produced, would be classified as hazardous, because hazardous waste is defined as waste that is (at least potentially) chemically active and if active is dangerous. According to official reckoning about four million tons of hazardous waste arise every year. The quantity would have been larger in the heyday of the alkali industry, though sulphuretted hydrogen is not as dangerous as some of the chemicals handled by modern waste contractors. Control of hazardous waste has been slow in coming. The alkali inspectorate took an interest when fumes might be given off. In 1952 for example they inspected nearly 500 spoilbanks, over half of which gave off no smoke or fumes; just over 200 gave off a little, and 26 gave off seriously obnoxious smoke or fumes. Other hazards such as the pollution of groundwater were outside the scope of the alkali acts, and only planning rules or the law of nuisance could be invoked. Positive control of such dangers only came with the Deposit of Poisonous Waste Act 1972 and with the Pollution Act of 1974. These acts regulated waste disposal on land by requiring contractors to have a licence. Well-disposed and efficient contractors subject to stringent inspection could hardly go wrong. On the other hand lax supervision of irresponsible contractors could have serious consequences. There has been much anxiety about toxic or hazardous waste since concern about pollution became widespread in the late 1960s. Undue quantities of nitrates have been detected in the water supplies of East Anglia but this problem arises from the heavy use of fertilisers in agriculture, not from the disposal of waste. Some land used for dumping toxic waste has no doubt been badly polluted, but whether by luck or good management groundwater has, so far, been little affected. Unless toxic waste does escape into aquifers the problem is essentially one of land loss on a comparatively small scale.

Derelict land

It might be thought that heavily polluted land would remain derelict for an indefinitely long time, but this is not necessarily the case. The Swansea Valley was a classic example of a landscape ruined by two hundred years of sulphurous fumes and the dumping of wastes from tin and copper smelting. In 1967 the *Lower Swansea Valley Project* published by University College, Swansea, aroused hopes that this notorious scene of industrial dereliction might yet be reclaimed for some useful purpose. If the Swansea Valley could recover there was hope for all the thousands of acres of industrial dereliction in South Wales – and the rest of the country. The proposals for the Swansea Valley had a sympathetic reception for they came in the year following the Aberfan disaster. Aberfan was a mining village south of Merthyr Tydfil: after heavy rain a colliery slag heap collapsed, overwhelming a local school in a sea of mud and rock; 116 of the 140 dead were village children. Land that had been an eyesore was now shown to be a danger as well. With the cooperation of the Nuffield Foundation and of the Forestry Commission, Swansea City Council set about reclaiming the 800 acres of derelict land in the Swansea Valley. Over twenty years later the work is nearing completion and only a small area of dereliction remains. An industrial estate set in parkland, woodlands together with several sports arenas have been developed. A further substantial area of land has been cleared but has yet to find a new occupier. Without generous grants from the Welsh Development Agency much less would have been achieved, for the costs of reclaiming land exceed the value it commands in competition with 'greenfield' sites for industry.

The Swansea Valley represents only a small part of the dereliction and of the movement for reclamation. After Aberfan the National Coal Board actively sought ways of clearing slag heaps or at least of making them safe, and since 1982 local authorities have been entitled to 100 per cent grants for reclaiming derelict land. It is no easy matter to discover how much land is derelict if only because the term is not readily definable with any accuracy. In 1954 it was estimated that 125,000 acres in England and Wales were derelict. Some of that has since been reclaimed, while other land has become derelict instead. Whether the land loss is regarded as large or small is a matter of taste: as a proportion of the land mass of England and Wales it is only one part in 300; on the other hand, it equals the area within an eight-mile radius of the Houses of Parliament, or half the area of Dartmoor National Park.[6]

New development

Derelict land is failed development and is of little importance compared to the land taken each year for new development. In medieval times, when the population of London was probably less than 50,000, the square mile of the City accommodated most of them with room for gardens as well. The encroachment of town on country has gone on remorselessly ever since, with short pauses for war, plague, or slump. Town populations grew most rapidly in the nineteenth century, but by the standards of modern suburban development laid claim to relatively little land. The City of London was particularly economical. In 1831 it had a population of 124,000; over half of London's population in that year occupied only 31 square miles, with a density of more than 33,000 per square mile.* Until after the First World War it was usual to provide only a minimal garden, back or front, except for the well-to-do in towns and for those who lived in the country where land was cheap. In south-west London near Clapham Common the big houses formerly owned by rich members of the evangelical movement were being demolished by the end of the nineteenth century, to be replaced by rows of solid terrace houses. If they have as much as 35 feet of garden the estate agent will mark it as a noteworthy feature when he draws up particulars for a sale. A front garden as much as 6 feet deep is a rarity. Yet these houses with their three or four bedrooms would then have been let and are now sold to prosperous members of the middle class. Suburban development since the First World War has made much more extravagant demands on land, with gardens back and front, wide roads, and occasionally grass verges as well as pavements. Industrial buildings too take up more room: the

* i.e. the City, Westminster, and the Holborn and Tower divisions of the county of Middlesex, with a population of 1,032,000.

six-storey cotton mill on a restricted site has given way to long low buildings set in more generous grounds. It is probable that motorways are greedier of land than the railways that they are replacing. Conventionally it is assumed that 50,000 acres are lost by agriculture every year; it follows that the whole of the land now under crops and grass will be covered by brick and concrete before AD 2600, at the present rate of loss. This estimate seems to come from evidence collected by the Ministry of Agriculture for the years 1926–36 (see Table 5.1). The Ministry's statisticians concluded that the average loss was 50,000 acres on the optimistic and unsubstantiated ground that some land reverts to agriculture each year. A better defence of this figure could be mounted: neither moorland nor allotments were truly lost. The allotments, intensively cultivated as they were by unpaid labour, added to rather than reduced the output of food and flowers. And moorland could always be reclaimed from the bracken at comparatively little cost if the need arose. Sports grounds too (and with more profit) could at a pinch be turned to good agricultural effect: their grassy pitches must store a great deal of locked-up fertility. Last but not least much of the land taken for building became gardens, and sometimes if not always grew useful crops. Whatever the true figure, it is clear that building and development take the lion's share of the land lost to agriculture each year. The loss in itself is small but can hardly be tolerated indefinitely.

Table 5.1 Agricultural land lost, 1926–36

	(000 acres)
Building and development	441
Reverted to moorland	192
Allotments	24
Waste	81
Sports grounds	105
Miscellaneous	41
TOTAL	884

LAND PRICES

So far, the encroachments on agricultural land have had remarkably little effect on the price of land. In city centres prices are measured in pounds per square foot rather than per acre, but in the countryside agricultural land has not fetched extravagant prices except in the early 1870s and immediately after the First World War. The price of land in the 1930s was once more low, so low that it was worth no more than in the 1890s, (a period of deep depression for agriculture) or indeed than at any time in the nineteenth century. In the past fifty years land has shared in the general inflation. Land that before 1939 might have sold for £20 or £30 an acre will now fetch about 100 times as much. House prices have advanced even faster, so that there appears to be nothing exceptional in the behaviour of land prices, and part of the increase may be a process of catching up, land prices having fallen further than other prices in the deflationary years 1921–35.[7] None of this accords well with the harsh and melodramatic vision of the classical economist, David Ricardo. In his view, rising population and a fixed supply of land could lead to only one result: stagnant wages and profits, together with exorbitant rents for the landlords who commanded a crucial factor of production. Britain's population has quadrupled since Ricardo's day – with none of the dire consequences that he predicted. The explanation is simple and well known. The acreage of land lost to houses, factories, schools, roads and railways has been regained many times over through the use of land overseas that has supplied Britain with food, industrial crops and minerals. The arm of the British economy has proved much longer than Ricardo or Malthus envisaged. There is no reason to doubt the soundness of Ricardo's analysis in the long run if and when the world comes up against the limits of food production. If Ricardo and Malthus could be consulted today they would probably express great surprise that the limits have yet to be reached in Britain. They cannot be blamed for not foreseeing the improvements in transport that have made the whole world Britain's farmyard, or the improvements in agriculture that enable British farmers to supply over half the food consumed in modern Britain.*

* Though Malthus was confident of the general truth of his system, he was cautious about making definite forecasts. But he did once venture to estimate how fast the population of the United States of America would grow. Although the USA was his classic case of rapid population growth, he thought it unlikely that it would attain the then population of China (some 330 million, he believed) in less than five or six hundred years. When he made this forecast the population of the USA stood at less than seven million. Today, not two hundred years later, it has already passed 250 million, and at recent rates of growth will reach 330 million within about 30 years. (Malthus, *Population,* 7th ed, 1872, p 258. The passage first appeared in the 1807 edition.)

REFERENCES

1. J A Steers, *The coastline of England and Wales* (Cambridge, 1976), pp 682–3; r.c. on coast erosion, *Third report* (PP 1911 XIV), pp 43–4, 79, 127–8, 144; Evidence (ibid) qq 15349–71; J A Steers, *The sea coast* (1969), pp 31–2; H C Darby, *The draining of the Fens* (Cambridge, 1956); M A Havinden, *The Somerset landscape* (1981), pp 165–7; *Report of departmental committee on coastal flooding* (PP 1953-4 XIII), pp 4, 9–13, 23–4.
2. K A H Murray, *Agriculture* (History of the Second World War, civil series ed K Hancock, 1955), pp 250, 257; *Agricultural statistics*, 1936, 1945–9, 1967, 1987; News bulletins, Radios 3 and 4, 4 January 1990; R D Hodges and C Arden Clarke, *Soil erosion in Britain* (Soil Association, Bristol,1986).
3. John Evelyn, *Silva* (1664, repr 1776 with notes by Alexander Hunter MD, 4th edn 1812), II, pp 217–8; O Rackham, *Trees and woodland in the British landscape* (1976), pp 82–3, 90–3, 96–7; *Agricultural returns for Great Britain* (1872, 1881, 1891, 1895); *Agricultural statistics 1906* (PP 1906 CXXXIII).
4. R.c. on coast erosion, *Second report* (PP 1909 XIV), pp 7–10, 35–8, 40–2; *Third report* (PP 1911 XIV) p 17; Committee on national expenditure, *Second report*, (PP 1922 IX), pp 53–4; [May] Committee on national expenditure, (PP 1930–1 XVI), p 130; *Parliamentary debates*, 1 March 1922, cols 443–4; Estimates committee, *Seventh report: the forestry commission* (PP 1963–4 V), pp 2–4 and qq 61 ff; Forestry Commission, *Reports*, v.d; H J Barnett and C Morse, *Scarcity and growth: the economics of natural resource availability* (Baltimore, 1963), pp 209–11; *Forestry in Britain: an interdepartmental cost/benefit study* (HMSO, 1972), p 38.
5. *Municipal year book for 1898* and *1928, 1929, 1980*); W Short, 'Air pollution control – incinerators' (NSCA, *51st Annual conference 1984*), p 1; W Robertson and C Porter, *Sanitary law and practice*, pp 106-15; D V Jackson and A R Tron, 'Energy from municipal wastes' (NSCA, *1985*), p 1; *New York Times*, 9 September 1988; Shanks & McEwan plc, *Transfer and disposal of local authority waste, case history no.1, Avon County Council*; Western Riverside Waste Authority (representing boroughs of inner West London), *Waste disposal plan* (typescript 1989); House of Lords committee on noxious vapours (PP 1862 XIV), qu 1725; r.c. on noxious vapours (PP 1878 XLIV), qq 5485, 5493; D W F Hardie, *A history of the chemical industry in Widnes*, (Liverpool 1950), p 106.
6. *89th report on alkali etc. works 1952* (1953), pp 6–7; *Lower Swansea Valley – legacy and future* (Swansea City Council, 1982); E M Bridges, *Healing the scars: derelict land in Wales* (Swansea, 1988).
7. *Agricultural statistics, 1936*, p 167; Norton, Trist & Gilbert, 'A century of land values' [1781–1880] repr in *Essays in economic history*, ed E Carus-Wilson, (1962), III pp 128–31; G R Porter, *Progress of the nation* (1912), ed F W Hirst, pp 699–701; G E Mingay, *The gentry* (1976), pp 172–3.

Overleaf:
'The letter killeth'. Conservation like any other good cause can be taken too far.

"They didn't want to destroy the charm of the village completely so they kept the old pub."

CHAPTER SIX

Towards a Green and Pleasant Land

In China, years are named after animals – the year of the rabbit, the year of the tiger. . . . We prefer to call our years after good causes. 1970 for example was European Conservation Year, and organisations committed to conservation gained welcome publicity as a result. The Council for the Protection (previously Preservation) of Rural England was one such beneficiary. Though it had fewer than 2,000 members and spent only £34,000 in the year,* it attracted nearly 1,200 notices in local and 250 in national newspapers, and was mentioned in radio or television broadcasts on average once a fortnight. If an unobtrusive and orderly body could get so much publicity, conservation had evidently become newsworthy – a position, broadly speaking, that it has maintained ever since. Conservation means keeping what is thought to be worth keeping – buildings, a townscape, mineral reserves, a landscape, rare plants or birds, peace and quiet – in fact anything that is valued and seems to be threatened by 'development' or economic change. The term also refers to concern about pollution of land, air and water, though strictly speaking in matters of pollution the restoration of a lost state of purity is meant; no champion of the environment would wish to 'conserve' pollution. 'Amenity' is another term popular in environmental circles. It gained currency in the language of town planning through mention in the Housing, Town Planning Act of 1909. The word, meaning originally pleasantness, derives ultimately from the Latin verb amare to love. It conveys a more constructive idea than conservation, for amenities can be created – a playing field from a rubbish tip, for example – whereas conservation refers only to the preservation of existing amenities, inherited from the past.

* So exceeding its income by £12,000!

A FEELING FOR THE PAST

A sense of history often inspires the conservationist, whether he knows it or not. It was obviously present when 'Caesar's Camp' on Wimbledon Common was destroyed in 1877, to the outrage of antiquarians. It is just as obvious in the recent opposition to a new road across the site of the battle of Naseby. It is less obvious when a botanist protests at threats to Oxleas Wood in south-east London, or to the habitat of a rare wild flower. But the botanist's concern is as historical as it is scientific: he is trying to preserve something that has come down to us from the distant past. Curiosity about the past and delight in old things and deeds is a common, perhaps almost universal, trait. The *Iliad* and the Old Testament show that the sense of the past was strong at the very beginning of recorded history. Anglo-Saxon England had its historians and chroniclers, and in the twelfth century Geoffrey of Monmouth wrote a persuasive romance, *The History of the Kings of Britain*, that masqueraded as history for several hundred years. Monks and pilgrims had long revered the relics of English saints, but more general antiquarian studies in England were unknown before the fifteenth century. There may well have been earlier antiquarians, but their names have not come down to us. Antiquarians are of particular interest in the history of the conservation movement because their study of monuments and other material remains often made them sensitive to the destruction or even the restoration of these mute witnesses to history.

John Rous (1411–91) and William of Worcester (1415–82) are among the earliest antiquarians known to modern scholars. Duke Humphrey of Gloucester, younger brother of Henry V, more sceptical than most students of his age, rejected the mythical origins of the English royal family as invented by Geoffrey of Monmouth, declining to believe that it was descended from Brutus, a Trojan prince who had allegedly settled in Britain. By the end of the seventeenth century when historians finally abandoned Geoffrey of Monmouth's baseless absurdities, antiquarian studies were already a flourishing branch of learning.

The disinterested pursuit of truth was, however, not the only motive that inspired them. John Leland (1506–52), the king's librarian and antiquary, undertook his itinerary of England and Wales at least in part to secure monastic books for the king's library. As he told the king in 1545 his intention was:

> That the monuments of ancient writers as well of other nations as of your own province might be brought out of deadly darkness to lively light. . . . First I have conserved many good authors, the which otherwise have been like to have perished . . . that all the world shall evidently perceive that no particular region may justly be more extolled than yours for true nobility and virtues at all points renowned.
>
> I was totally inflamed with a love to see thoroughly all those parts of this your opulent and ample realm, that I had read of in the aforesaid writers: insomuch that all my other occupations intermitted I have so travelled in your dominions both by the sea coasts and the middle parts, sparing neither labour nor costs, by the space of these vi years past, that there is almost neither cape, nor bay, haven, creek or pier, river or confluence of rivers, breeches, washes, lakes, meres, fenny waters, mountains, valleys, moors, heaths, forests, woods, cities, boroughs, castles, principal manor places, monasteries, and colleges, but I have seen them; and noted in so doing a whole world of things very memorable . . .
>
> I trust so to open this window that . . . the old glory of your renowned Britain [may] reflourish throughout the world. (*Leland's Itinerary*, I, pp xxxvii–viii.)

Leland intended to write a work called 'De Antiquitate Britannica' or 'Civilis Historia', county by county, and it was for this purpose that he compiled the rough travel notes that we call *Leland's Itinerary*. He never wrote the ambitious book or books that he had planned, and his notes seem too scrappy and miscellaneous to form the basis of guide books, let alone county histories. He had a wide range of interests, but his comments were usually short and not always informative. His description of the parish church of Glastonbury is longer than most and may stand as an above average example of his powers of observation:

> This is a very fair and lightsome church: and the east part of it is very elegant and aisled. The body of the church has . . . arches on each side. The choir hath three arches on each side. The quadrate tower for bells at the west end is very high and fair. (Ibid., p 148.)

He paid particular attention to bridges and fortifications, noted economic decay (above 200 ruinous houses at Bridgwater), the state of the roads, and the appearance of the countryside – woodland, grass, and arable. *Leland's Itinerary* is a quarry for historians to this day, not for its intellectual distinction, rather because it is one of very few topographical sources at so early a date.

Leland's patron, the king, was no conservationist: some monastic books he may have coveted, but the dissolution of the monasteries led to the wholesale destruction of some of the country's finest buildings: the ruins of Tintern, Glastonbury and Fountains give some idea of the riches that have perished. When greed had done its worst, Protestant fervour continued the assault on ancient religious monuments. Elizabeth I and her Council were anxious to stem the tide of destruction. For example they took a sympathetic interest in the condition of the Eleanor Cross in West Cheap in the City of London. Erected by Edward I in 1290 it was one of twelve elaborate crosses that marked the journey of his dead Queen Eleanor to her final resting place. Restored in 1441 the Cross was an obstruction to traffic by the reign of Elizabeth. In 1581 its images of the Virgin Mary were defaced and by 1599 the monument was headless. The Privy Council ordered its repair, referring to 'the antiquity and continuance of that monument, an ancient ensign of Christianity'. A year later the cross was still encased in scaffolding; a sharper message from the Council to the city fathers led to a speedy repair, but within a fortnight religious vandals had defaced it once more.[1]

A similar fate overtook many statues and much stained glass in the seventeenth century. And the west front of the old St Paul's was encased in wholly inappropriate classical columns by Inigo Jones, not to mark his antipathy to 'Romish superstition' but to express his preference for modern Italian over medieval English architecture. Wren and Hawksmoor were more tolerant and, where tact demanded it, were prepared to design in the Gothic style, notably at Christ Church and All Souls, Oxford. James Gibb did not feel the same compunction when he built the Senate House in Cambridge close by King's College Chapel. And it is fair to regard the leading architects in the eighteenth century as wedded to classical styles unless bored or whimsical clients decreed otherwise. Chinese, Hindoo, Etruscan, Greek and Gothic(k) all attracted some attention, but Gothic, the only style with historical roots in England, eventually triumphed, for a time, in this drawn-out contest. Victory was assured when the Barry-Pugin designs won the competition for the new Houses of Parliament in 1836. Only a few years before Greek Revival had been the chosen style for the British Museum.

The renewed interest in Gothic led to research into the chronology of medieval architecture and to a genuine understanding of the style. Nobody could mistake Strawberry Hill for a medieval building, but many Victorian churches closely followed medieval practice, and some at least caught the spirit of the age of the faith. Criticism of Victorian architects has been directed less at their new churches, though they were long out of favour, than at their repair and restoration of medieval churches. Though others had already voiced concern, it was William Morris, poet craftsman and socialist, who set in motion the English movement for the conservative preservation of historic buildings. In 1877 he wrote a letter to *The Athenaeum* protesting at recent restoration practice and suggesting the formation of a society:

> My eye just now caught the word 'restoration' in the morning paper, and on looking closer I saw that this time it is nothing less than the Minster of Tewkesbury that is to be destroyed by Sir Gilbert Scott. Is it altogether too late to to do something to save it – it and whatever else of beautiful and historical is still left to us on the sites of the ancient buildings we were once so famous for? Would it not be of some use once for all, and with the least delay possible, to set on foot an association for the purpose of watching over and protecting these relics, which scanty as they are now become, are still wonderful treasures, all the more priceless in this age of the world, when the newly invented study of history is the chief joy of so many of our lives?
>
> What I wish for therefore is that an association should be set on foot to keep a watch on old monuments, to protest against all 'restoration' that means more than keeping out wind and weather, and by all means, literary and other, to awaken a feeling that our ancient buildings are not mere ecclesiastical toys, but sacred monuments of the nation's growth and hope. (Mackail, *Life of William Morris,* I, pp 351–2.)

This trenchant appeal led to the formation of the Society for the Protection of Ancient Buildings, or Anti-Scrape, as Morris dubbed it. One of the society's first tasks was to try to prevent the restoration of the west front of St Alban's Cathedral: 'We could scarcely point', wrote Morris 'to any fragment of ancient architecture we so much desire to keep unaltered, even to its latest detail.'

Admitting that in its decay the west front was less beautiful than it once had been, Morris nevertheless argued for its preservation because it still had vestiges of Abbot John de Cella's design in it, and because of its pathetic appeal despite mutilation and defacement. He protested against the restorer's ideals of neatness and smoothness 'the virtues that rank highest just now in the popular estimation of works of art'. Restoration to Morris was dishonest because it introduced modern work masquerading as old; it was also dull and mechanical since, he believed, modern craftsmen lacked the skill and idealism of the medieval mason. There should be no pretence of fine art, no restoration, no decay.

Victorian England had all the engineering skill that ancient buildings needed 'and this is all they ask at our hands. . . . Preservation in every sense of the word; this is our duty to them and to their builders'. (SBAB, *St Alban's,* 1879, p 2.) The preservationists' austere creed did not deflect Sir Edmund Beckett, later Lord Grimthorpe, from saving St Alban's in his own way; like other restored churches St Alban's remains an impressive monument and nobody else seemed able or willing to supply the huge sums necessary to save it.* Despite Morris, churches are not primarily museums or fragments of art history, but places of worship. It might satisfy the preservationist to patch Westminster Abbey with purple stocks rather than deceive with modern stonework; he has cheerfully left featureless medieval sculpture in the screen of the west front of Exeter Cathedral rather than replace it with clean-cut modern work. But if this painful honesty affronts worshippers, the building no longer fulfils its primary purpose; if it affronts tourists as well, the stern pleasure of a few perhaps mistaken experts has to outweigh a great deal of disappointment.

CONSERVATION OF BUILDINGS AND MONUMENTS

In its first year 'Anti-Scrape' considered forty threatened buildings; ninety years later its workload had multiplied tenfold, not because more buildings were in danger but because more people were concerned about the danger. It had also assisted at the birth of other conservationist bodies – the National Trust (1894), the Georgian Group (1937), the Civic Trust (1957) and the Victorian Society (1958). It encourages the new discipline of industrial archaeology and has a section devoted to the preservation of wind and water mills. After the Second World War civic societies sprang up in many towns to do locally what the Society for the Protection of Ancient Buildings and other national bodies tried to do on a larger stage. When in 1947, all development became subject to planning permission, from the local authority in the first instance, civic societies were well placed to scrutinise planning applications,

* 'One cannot deny that there is something to show for the money' is the grudging admission of the royal commission on historical monuments (1982).

comment on 'road improvements', and develop a conservationist strategy. Until town planning became a moderately effective instrument of government the influence of conservationists had been limited. It is easier to sway an elected local authority than a company or the private owner of a threatened building.

Modern planning law gives conservationists many opportunities to oppose development, and it is instructive to compare the building of the Great Western Railway from London to Bristol with the building of the M4 motorway between the same two cities. Opposition to the railway's route had to be expressed within the private bill procedure of parliament. The bill authorising the Great Western Railway passed through all its stages in the 1834–5 session, and, once authorised, construction went ahead with great speed so that the whole length of railway was open by 1841. Planning approval for the M4 motorway was sought piecemeal, section by section, over a period of 14 years; resistance was eventually overcome, but opponents of the motorway at least had the satisfaction of fighting a long war of attrition and they were not defeated in one set-piece battle as the railway's opponents had been in 1834–5.

Voluntary associations and private persons can do much, but their most effective role lies in persuading parliament and government to control development in the interests of conservation, and to protect monuments that may be threatened either by decay or by development. The term 'ancient monument' was first recorded in the year 1525. It had no precise meaning but tended to refer in the late nineteenth century to uninhabited secular and often prehistoric structures. It is in this sense that the term is used in the Ancient Monuments Protection Act, 1882. Sir John Lubbock first introduced an Ancient Monuments Bill in 1876, a year before the foundation of the SPAB, but it was not until 1882 that a measure sponsored by government reached the statute book. The act empowered the Office of Works (now hideously renamed the Property Services Agency) to accept guardianship of ancient monuments and to maintain and fence them at public expense. There was provision for the purchase of monuments – it was more than 30 years before this power was exercised – but in this first tentative measure owners were under no obligation either to sell or place in guardianship the monuments in their possession. Nor were they obliged to preserve them in any other way. The act listed some ancient monuments, mostly prehistoric, and provided for others to be added to the list. It also provided for the appointment of one or more inspectors. In 1900, county councils were given similar powers of maintenance and purchase, and public access to monuments in official care was guaranteed. By 1912 the Office of Works had charge of more than 100 ancient monuments besides another 30 belonging to the War Office; these included the Tower of London and Edinburgh Castle.

Most of the monuments came to the Office of Works after 1900; had owners been better informed about the law it is likely that more monuments would have been handed over. Tattershall Castle, dating from the mid-fifteenth century, was not among them. Its owner, alert to the possibilities, stood out for a 'fancy price' from the Office of Works, which refused to pay and was therefore powerless to save the building. Lord Curzon, acting privately thereupon, bought back its carved gothic fireplaces, which had reached London on their way to America. He replaced them in the castle, restored it and in 1925 bequeathed it – and Bodiam Castle – to the National Trust.

The Tattershall Castle affair provoked an enquiry and a change in the law. A new act passed in 1913 deliberately excluded churches and houses from its scope; only unoccupied buildings and other remains were eligible for the protection that the law now offered. The unfettered rights of private owners to allow their monuments to decay and fall into ruin came to an end. The Office of Works was empowered to make a preservation order taking threatened monuments into guardianship, subject to parliamentary approval. For the better preservation of the country's heritage it was important to ascertain what that heritage was. The inspector of ancient monuments was therefore to prepare a list of such monuments under the guidance of an Ancient Monuments Board. A further amending act in 1931 required the Office of Works to serve a notice on the owner of a listed monument; this notice had the effect of preventing alteration or demolition unless three months notice had been given. In three months the Office could make a preservation order if it so wished.

Ancient monuments as defined by law include only a tiny fraction of the country's architectural heritage, though much more of its archaeological remains. Houses, churches and other buildings still in use remained at the mercy of their owners. In 1907 a standing royal commission on historical monuments was set up to record monuments of a date earlier than 1700 – a time limit since removed. Monuments went undefined but they had to be 'connected with or illustrative of the contemporary culture, civilisation and conditions of the people in England'. The commission was to indicate which monuments were most worthy to be preserved, even though there was no machinery for their preservation unless they came into the narrow class of 'ancient' as distinct from 'historical' monuments. The commission has proceeded with scholarly deliberation: by 1950, 9 English, 13 Scottish and 8 Welsh counties had been surveyed in 40 handsome volumes. Dry, precise and comprehensive, the reports amount to a thorough inventory of structures and buildings, but with few aesthetic judgements and no overall impression. The reader of the account of Amersham in south Buckinghamshire, for example, will find 68 buildings listed and described but he is not told that Amersham is one of the best preserved and most attractive small towns in England. A more popular survey of historic architecture is to be found in Pevsner's *Buildings of England* published in many volumes since the Second World War. Warm, personal, selective and hasty, Pevsner stands at the opposite pole from the royal commission, though they meet on common ground in bearing witness to the public's interest in old buildings.

In 1941, the National Buildings Record began to compile a photographic record of as many old buildings as possible in case they were destroyed by bombing. There followed a more systematic attempt to list the country's worthwhile old buildings, under the Town and Country Planning Act, 1947: outstanding buildings were listed as grade I, others as grade II or grade III. In 1970 grade III was abolished, and some of the buildings in it were upgraded to grade II, while others were struck off. The criteria for listing have changed over time. Originally the lists were confined to buildings dating from 1840 or earlier, but distinguished recent buildings are now included, as are industrial monuments and buildings illustrative of the history of building technology. The number of listed buildings has grown very large: over 400,000 in 1990, of which some 6,000 were in grade I and a further 20,000 were grade II starred, as being of high quality without being quite worthy of grade I. Though listed status means that buildings may not be altered or demolished without the consent of the local planning authority (or, on appeal, of the minister) listing does not necessarily guarantee preservation. In 1981 for example when there were over 280,000 listed buildings, 147 were demolished, none of them in grade I and only eight grade II starred; ten times as many (1418) were partly demolished in grades II and II starred. The only demolition at grade I sites was of outbuildings.[2] Whether the loss of one listed building in 2,000 in one year would amount to a serious attack on the country's architectural heritage seems doubtful. Indeed the reported rate of destruction is so low as to arouse suspicion, and it is to be feared that the Historic Buildings Council, the source of these figures, can scarcely have had full information on the fate of all listed buildings. If the figure were correct, listed buildings despite their age fared six times as well as houses in general, for of the total housing stock about three per thousand were lost in the course of 1981. It is only fair to add that 1981 was a peak year for slum clearance, and that on a ten-year average fewer than two houses per thousand are lost from the general stock. On the face of it, the procedure for listing and preserving individual historic buildings is working well.

An attractive old village, town, or part of a town does not necessarily rely for its charm on a few buildings of special merit; it is often a combination of individually modest but collectively picturesque or stylish buildings that gives an area its appeal. In these cases it is not just a few buildings of high quality but a whole area that calls for conservation. In Britain the classic example is the largely Georgian city of Bath. Blackened by soot and suffering from many years of neglect, the buildings of Bath were in a poor state of repair by the end of the Second World War; the claims of the motor-car and the lorry threatened to add to the toll of destruction, and it was not until 1973 that plans for a motorway through the heart of the city were scotched. By that date many Georgian houses had been lost through demolition and redevelopment. Many more however had already been cleaned and restored to their former beauty, and more have since been rescued, so that Bath remains to this day a largely Georgian city, flourishing and prosperous. Bath is a good example of conservation because in it usefulness goes hand in hand with beauty. William Morris would have approved – had he admired Georgian architecture.

Where Bath led the way in the 1950s other historic towns gladly followed in the next decade, notably York, Cheltenham and Stamford. By 1972, 31 schemes had been drawn up for conservation areas in historic towns. In that year an amending act authorised grants in aid of the restoration of such areas. Within ten years local authorities had designated over 5,000 conservation areas: conservation, once the scholarly interest of a few, now commanded widespread public support. Householders no longer ripped out old fireplaces; rather they scoured antique shops for picturesque and not always authentic replacements, for old light fittings, decorative tiles and brass taps for the bath. For those with less money new mock-Georgian doors and gas or electric fires in the Adam style were readily available. The house with 'original features' in a conservation area invariably got a special mention in estate agents' publicity material. At the height of the housing boom in 1987–88 it was confidently asserted that the presence of original features might add as much as ten per cent to the value of a property.

The Ancient Monuments Acts had deliberately excluded churches and inhabited houses, where most of the country's architectural splendours were to be found. Before 1977 churches received no grants from the Historic Buildings Council, the appropriate government agency until the establishment of English Heritage in 1982. English Heritage is the officially sanctioned popular name for the Historic Buildings and Monuments Commission for England. It brings under one, more commercial, management three tasks formerly distinct: description (the Royal Commission on Historical Monuments); grants (the Historic Buildings Council); and the guardianship of monuments (the Ministry of Works). In the five years that remained to it the Historic Buildings Council made over 400 grants to churches, nearly all of them Anglican. That most of the £9 million of grants went to the Church of England should cause no surprise: it reflected not a high-tory plot but the rich architectural endowment of the established church. Grants to the owners of country houses had a longer history, and resulted from recommendations made by the Gowers Committee in 1950. The Labour government in power at the time kept taxes at such high levels that many owners of country houses could no longer afford to live in them. The government did not wish its pursuit of equality to result in the destruction of a valuable part of the country's architectural heritage, and the Gowers Committee proposed that government should give back with its left hand part of what it had taken away with its right: owners of country houses of historic and architectural interest, it was proposed, could apply for grants for maintenance and repair; in return their houses should be opened to the public for a reasonable number of days each year. In 1953 the Historic Buildings Council was created to administer the grants. In its first year it disbursed a quarter of a million pounds; in its last, £35 million. At the latter date (1981–2) it was administering grants to conservation areas as well as to churches and country houses, and was encouraging the conservation of furniture and furnishings, and the restoration of historic gardens. The sense of the past had seemingly never been stronger.[3]

ACCESS TO THE COUNTRYSIDE

A sense of history was not the only inspiration for conservationist effort. Concern for the preservation of peace and quiet and natural beauty has also been a powerful force. Often selfish motives have been evident – in Wordsworth's opposition to railways in the Lake District, and recently in the protests against reservoirs or roads on some of the most barren and unlovely stretches of Dartmoor. Economists have pointed out that shared pleasures – motoring on a bank holiday, sunbathing on a crowded Brighton beach, or looking at the Turners while being jostled by tourists in the Clore Gallery – may be no pleasure at all. In so doing they almost make a respectable case for those who, like the Dartmoor Preservation Society, take the 'not-in-my-backyard' stance. For the most part however those who have sought to preserve natural beauty have done so hoping to share their pleasure with others, and to encourage public health through access to the open air in enjoyable surroundings. It was in this fashion that the best known and most powerful of Britain's conservation bodies began – the National Trust. The Trust was by no means the first body interested in securing open spaces for the people's enjoyment. Peel Park in Salford and Victoria Park in East London were created in the 1840s by a mixture of philanthropy and public enterprise. The Crown developed Battersea Park in the 1850s out of marshland on the banks of the Thames. From the 1860s the Metropolitan Gardens Assocation began turning small disused London churchyards into public gardens, a work especially useful in East London, which was sadly lacking in open spaces. It was in the 1860s too that the City of London stirred itself to preserve 3,000 acres of Epping Forest from developers.

Allied to the City in this campaign was the Commons Preservation Society; and their young solicitor Robert Hunter (1844–1913) was later prominent among the founders of the National Trust. Sir Robert Hunter, as he had then become, lived at Haslemere and it is perhaps no accident that the Devil's Punchbowl and other land at Hindhead near Haslemere were among the Trust's earlier acquisitions. Octavia Hill, another founder member of the National Trust, had jointly set up the Kyrle Society in 1877; one of the aims of the Society was to secure open spaces in and near London. Another society that had much in common with the National Trust was the Society for Checking the Abuses of Public Advertising [SCAPA] founded in 1893. Its council included many distinguished figures – architects, poets, artists, lawyers, scientists, economists and historians among them. The ordinary members included Robert Hunter and Canon H D Rawnsley. SCAPA published the first number of its journal *A Beautiful World* in November 1893; Rawnsley, as acting secretary of the embryonic National Trust, took the endpaper to advertise the Trust and to appeal to the owners of places of natural beauty. They might not care, he recognised, to give land to their local authority even if there were one willing to assume the responsibility. The National Trust, he hoped, might be an acceptable alternative. Or it might encourage owners to enter into restrictive covenants, which would have the effect of preserving places of natural beauty. In time, he thought, the Trust might accumulate funds and buy outstanding properties itself.

In this modest way the National Trust came into public notice. It was formally constituted in 1894 with Rawnsley as secretary. Well connected from the start – representatives of bodies as various as the National Gallery, the ancient universities, the Society of Antiquaries and the Selborne Society sat on its council – it had fewer than 150 members in 1896 and a tiny income. There was no fixed subscription, most members paid a pound or a guinea, some as little as five shillings. When a building of historic or architectural interest came on the market, or an area of natural beauty, the Trust appealed for funds – as it still does. If it had had £250 it could have bought the cottage at Nether Stowey in the Quantocks, where Coleridge wrote *Christabel* and *The Ancient Mariner,* but it came to the Trust – by gift – only in 1909. The first purchase, the Clergyhouse at Alfriston in Sussex, cost only £10; however, its restoration under the advice of the SPAB was much more expensive at £234 12s. 9½d. The first purchase of land, 14 acres at Barras Head, near Tintagel, was made in 1897; the Trust had already been given 4.5 acres of cliff-top near Barmouth in 1895. On the eve of the First World War, when membership had still not reached one thousand, the Trust was a substantial landowner with more than 60 properties and 6,000 acres. Its acquisitions included land at Derwentwater and Windermere in the Lake District;* Wicken Fen in Cambridgeshire; Cheddar Cliffs in Somerset; Blakeney Point in Norfolk; and Box Hill, a popular beauty spot in Surrey. In addition the Trust owned a few buildings, picturesque, old, or with literary

* Canon Rawnsley, the Trust's secretary, was vicar of Crosthwaite near Keswick from 1883 to 1917.

associations – among them the Old Post Office, Tintagel. Its only substantial country house was Barrington Court in Somerset.[4]

Between the wars the National Trust prospered and grew to be a considerable landowner. It had several thousand members by 1938, an income of nearly £10,000, and 45,000 acres of land, mostly in areas of natural beauty. In addition the Trust held covenants restricting development over a further 16,000 acres. The list of interesting or beautiful houses remained short. Among them were Montacute in Somerset, Quebec House in Westerham, Kent, and the Treasurer's House, York. In 1937 the Trust secured a new act of parliament that heralded a change in direction. The 1907 act empowered the Trust to acquire buildings of beauty and historic interest; in the new act the wording changed to 'buildings of national interest or of architectural, historic or artistic interest or beauty'. The new wording allowed the development of an interest in the history of taste and encouraged the Trust to acquire such properties as Waddesdon and Kingston Lacy that might have been excluded under the old, more restrictive wording.

A more important purpose of the act was to smooth the transfer of historic country houses to the Trust's ownership. Under its terms owners might surrender their property to the National Trust so avoiding death duties; in return the family would continue to live in the ancestral home on condition that public access to the principal rooms was allowed for a reasonable number of days each year. In the short period that remained before the outbreak of war in 1939 the scheme attracted few takers. Most owners of historic houses evidently felt that income tax and surtax reaching 75 per cent at the highest level were not intolerably burdensome. During the Second World War the highest combined rates of income and surtax came to exceed 90 per cent; they fell briefly below that level in the early 1960s, but stood at the very high level of 98 per cent (on unearned income) in the 1970s. Decisive reductions in the weight of direct taxation on large incomes began only in the fiscal year 1979–80. Under a severe tax regime lasting nearly forty years owners of country houses proved much more willing to transfer their properties to the National Trust. By 1950 the scheme applied to 50 houses, and in 1957 the Trust was opening more than 130 historic buildings to the public, four times as many as in 1936. The number of houses opened doubled again in the next thirty years, and the estate grew to some 600,000 acres.

Even more startling was the rise in membership: from less than 7,000 in 1936, to 65,000 in 1957, and in 1987 to 1.5 million! The growth of the Trust's property, income and membership greatly widened the range of its conservation work: it now includes the tending of great gardens like Bodnant, Sissinghurst and Hidcote; the renovation of textiles and furniture; the preservation of Cornish beam engines, and domestic archaeology – the restoration of Calke Abbey, a country house where virtually nothing had changed for sixty years. The Trust's original purpose – the preservation of areas of natural beauty in the interests of public health – is now only one object among many. Enterprise Neptune, an ambitious attempt to acquire the remaining 900 miles of unspoilt coastline in England and Wales, was launched in 1965. At that time the Trust already owned 175 miles of coast. It has since acquired by gift or purchase a third of the remaining unspoilt coast and now holds about half of what it reckons to be worth preserving. Enterprise Neptune is the best proof that the Trust has not forgotten the purpose that inspired its founders. When interest in the past is strong and half the families in the country own at least one car it would be unreasonable to confine the National Trust simply to the acquisition of places of natural beauty. There seems at present no danger that its original purpose will be forgotten in the recent headlong growth.[5]

No other voluntary conservationist body has enjoyed anything like the success of the National Trust. The SPAB and the Council for the Protection of Rural England (CPRE) remain small, specialised and quietly influential bodies. A few private owners of great country houses – the Dukes of Marlborough and Bedford and the Marquess of Bath – have attracted visitors on a scale that even the National Trust might envy, but the attractions offered have not always been scholarly or historical. Folk museums and industrial archaeology – areas of antiquarian interest only lightly touched upon by the Trust – have drawn much public support in recent years. In 1970, folk museums like St Fagan's near Cardiff or the Museum of Rural Life at Reading were rare in Britain, though already popular in Scandinavia; and the study of industrial archaeology had yet to attract much attention. Since then scarcely a wind or water mill survives that has not been pressed into service as a museum or at least a craft centre. Important new museums or parks of industrial archaeology have been established at Ironbridge and at Beamish in County Durham. The public appetite for domestic and industrial bygones seems insatiable. In the preservation of open spaces, and securing public access to the countryside the National Trust, though not alone among voluntary bodies, stands apart by the extent of its achievement. More specialised bodies like the Royal Society for the Protection of Birds and the Wildfowl Trust acquire land not so much to encourage public access as to create reserves where threatened species can breed undisturbed.

Much that has been done to preserve the countryside and to encourage public access to places of natural beauty has been beyond the power of voluntary bodies and has depended on changes in the law. Proposals for a 'green girdle' around London go back a hundred years, but nothing effective was done until the London County Council began to acquire land in 1935. Acting in cooperation with neighbouring counties that had their own reasons for stemming the onrushing tide of development, some 40 square miles of land had been set aside as 'green belt' by the outbreak of war. Some of the land had been bought outright, some subjected to restrictive covenants. In the absence of effective planning laws that was the only way in which local authorities could check ribbon development, prevent adjacent settlements from merging into one undifferentiated mass of building, protect farmland from development, or provide open spaces for recreation. The Town and Country Planning Act, 1947, drastically changed the law by subjecting all development to local authority approval, and by requiring the planning authorities to prepare development plans for submission to the minister. Proposals for a green belt round London had been put before the public for many years, most recently in Patrick Abercrombie's *Greater London Plan* published in 1945, and when the development plans came up for consideration between 1954 and 1958 a green belt some eight to ten miles wide round the built-up area of London was incorporated into them and received the minister's approval. A planning blueprint cannot stop development dead in its tracks, but the existence of an authorised green belt has checked ribbon development and has tended to divert new building to the region beyond the green belt. Within the green belt itself agriculture accounts for 70 per cent of land use, naturally with little public access. Woodland and recreational uses account for a further eighteen per cent of the land. Some of the woodland is private and much of the recreational land is devoted to golf courses, but there remain large tracts of land, much of it of great natural beauty, to which Londoners have access – if they own a motor car or a stout pair of walking boots.

The need for a green belt round London was strongly felt because of the great size of the London conurbation and the lack of open space – especially noticeable in east and north-east London. Other great towns were less cut off from the countryside, but in them too planners showed interest in the green-belt idea. Some – Birmingham, Sheffield and Leeds – imitated London and acquired land for open space in a green belt before 1939. When effective town planning became possible after 1947 several large towns took the opportunity to provide for a green belt round their built-up area and no less than a seventh of England has now been declared green belt. Unlike London, many great towns lay comparatively close to wild open country, sometimes bleak and desolate, sometimes of great beauty. The small body of ramblers and the much larger number of less energetic trippers and holiday-makers had an interest in access to and preventing development in the wilder parts of Britain. Those who lived on Dartmoor, near Snowdon, or in the Lake District might not share the townsman's enthusiasm for unrestricted access, but they were just as anxious to preserve the amenities of the area by preventing development.

These considerations lay behind the movement for national parks. In the United States and other thinly peopled countries the obstacles were fewer. The first such park, Yellowstone, was instituted in 1872, and in 1916 the United States went further and established the National Parks Service 'to conserve the scenery and the natural and historic objects and the wildlife therein'. In 1885 Canada created Banff National Park; in 1898 Kruger Park in the Transvaal became a game reserve; and by 1907 Argentina had no fewer than five national parks. The Belgian Congo, Australia and New Zealand, and several countries in Europe also established parks. The idea came under discussion in Britain in the 1920s but nothing was done because of the economic depression of 1929–33. At the end of the Second World War the project was taken up again with the backing of Labour's Chancellor of the Exchequer, Hugh Dalton, himself an enthusiastic rambler. Legislation in 1949 provided for a National Parks Commission (now the Countryside Commission) and for the creation of ten national parks in England and Wales: the Peak District, the Lake District, Snowdonia, Dartmoor, the Pembrokeshire Coast, the North Yorkshire Moors, the Yorkshire Dales, Exmoor, Northumberland, and the Brecon Beacons. The South Downs and the Norfolk Broads, both much closer to London, were excluded from the list and it was only in 1989 that a Broads Authority came into existence with powers of management similar to those exercised in the national parks. The national parks cover nine per cent of the area of England and Wales, so that, with the green belt, more than a fifth of the countryside now has stringent protection against development. In addition, the Countryside Commission and the Nature Conservancy Council now designate areas of outstanding natural beauty and sites of special scientific interest [SSSI], both subject to strict planning controls. The areas so named are extensive but sometimes overlap with green belt or national park: the North Downs are a case in point. This protective apparatus began with the National Parks Act, 1949, which marks the start of effective national concern with these matters.[6]

NOISE POLLUTION

Conservation and the prevention of pollution are problems of great antiquity to which few people gave much thought until comparatively recently. Other problems, understandably, seemed more important. In a Victorian philanthropist, let alone an interested party, poverty, the corn laws, child labour, or widespread drunkenness, would have aroused more anger than smoke, smells, or the loss of old buildings or picturesque scenery. An evil less considerable than any of these was the problem of noise, a problem which few could be persuaded to take seriously until the last forty years. Noise was recognised as a cause of deafness among some industrial workers quite early in the nineteenth century. Boiler-makers, smiths and braziers, weavers, railway workers and heavy forge operators all risked some loss of hearing. In more recent times aviation deafness has been added to the list. Many industrial and building processes are noisy and it is probable that some loss of hearing among workers is widespread, but there is no danger to life and occupational deafness is a comparatively minor concern, even to students of occupational diseases. Similarly, noise as loss of amenity aroused little interest, and complaints were few. A writer on the trades of London noted in 1747 that coppersmiths were 'very noisy neighbours and they ought to live by themselves'. The clatter of iron-shod wheels on cobbles made the streets of Victorian towns noisy, perhaps as noisy as streets are today. The great Whig historian Macaulay, no enemy of progress, found towns little to his taste as he grew older:

> Who can bear the thought of those labyrinths of brick, overhung by clouds of smoke and roaring like Niagara with the wheels of drays, chariots, flies, cabs and omnibuses? Give me rural seclusion like this [at Malvern] especially when I can smell, as I now do, a leg of mutton on the spit for my dinner . . . (*Selected Letters of Thomas Babington Macaulay*, p 248.)

The Metropolitan Police Act, 1839, outlawed street cries and the blowing of horns and other nefarious activities like gambling in the streets, sliding on ice, and flying kites to the annoyance of passers-by. Such grandmotherly legislation usually fails because the offences are so widespread and individually so trivial that constables look for other fish to net. Good manners and consideration for others are not so easily to be imposed by act of parliament on an unruly population, and it is still the case that persons as distinct from companies are virtually free to make as much noise as they please.

None of the noise nuisances so far mentioned was a product of the machine age: all could have been heard in Tudor, or indeed Norman, London. The use of elaborate power-driven machinery and a multitude of inventions based on scientific discovery have greatly added to the volume of noise as nuisance, as well as to noise in the workplace. Until trading estates became popular and town planning empowered local authorities to control the location of factories it was usual for industry and homes to lie close together. Heavy industry in a steel town like Sheffield could make life a misery for whole streets and, in the absence of other powers the city council condemned the affected houses as slums and rehoused the occupants elsewhere. Machinery and clanging metal were not the only sources of factory noise. Steam whistles summoned the workman to the mill and told him when to go home. The principle of the alarm clock had long been known and by the 1850s large numbers of Swiss and French clocks and watches were being imported to supplement or displace home production.[7] Yet the din went on and became so intolerable that in 1872 Parliament passed a short act forbidding the use of steam whistles without the consent of the local sanitary authority. When whistles were banned, sirens took over the task of announcing the beginning and end of the working day – and continued in use until the Second World War when they too were silenced except to give air-raid warnings and to sound the all-clear. The building trades had always been noisy: sawing, hammering, whistling at pretty women, the crash of falling masonry in demolition work, were none of them peculiar to modern times.

Invention has added new and louder forms of annoyance: the pneumatic drill, the pile-driver, the concrete mixer and the clatter of steel scaffolding (wooden scaffolding poles were a common sight until fifty years ago). Mercifully, neither factory nor building noise was all-pervasive, but a localised nuisance.

It was, and is, much harder to escape from domestic and traffic noise. Late-night parties, dogs barking, voices raised in anger are nothing new; the machine age has simply armed the inconsiderate with extra weapons. The iron-framed piano was the first of a long line of technical marvels that give neighbours the opportunity to annoy each other, either deliberately or out of carelessness. Radio, television and the gramophone, the latter now equipped with powerful amplifiers and two or even four aptly named 'loud-speakers', are the commonest sources of mechanical noise. The transistor or portable radio was no more than a prospective threat according to the Wilson committee on noise in the early 1960s. It had its day and has been succeeded by the less intrusive but still troublesome 'personal stereo'. The kitchen has become the second noisiest room in the modern house. Washing-machines and dish-washers, food-mixers, grinders and juicers, extractor fans and occasionally waste-disposal units combine to give it the air of a miniature factory. The family suffers most from the disturbance of the peace, but in flats and terrace houses neighbours often suffer as well.

Traffic makes louder noises than the most powerful domestic equipment. The railway has always been an exceptionally noisy form of transport. At a given speed it is probable that a steam train running on old-fashioned track made more noise than an electric train now does on continuously welded rails. However train speeds are often higher with modern equipment, and noise inevitably increases with speed. But since the highest speeds are reached in open country where fewer people live, that cause of annoyance affects relatively small numbers. The greatest weight of railway traffic occurs where lines converge on the approach to city centres and many thousands of homes lie within earshot; even at the somewhat lower speeds prevalent on congested lines the noise nuisance is considerable for the majority of people living nearby. Road traffic affects still more people.* For over thirty years, more freight has gone by road than by rail and the number of cars has

* Road transport first carried more freight than the railways in 1958; by 1970 roads carried three times as much freight as railways, and by 1986 six times as much.

remorselessly increased: two million in 1938; four million in 1957; eight million in 1964; sixteen million in 1984. Engineers have had some success in reducing the volume of noise made by individual cars and lorries, and the low speeds at which traffic moves in congested towns also keep noise levels below what they might be in a world of urban motorways. Nevertheless, road traffic noise is a serious and increasingly resented evil. Aircraft noise has a shorter and more turbulent history. In 1938 fewer than a hundred passenger-carrying aircraft left Britain on overseas flights each day. By 1970 the number of overseas flights reached a thousand a day, and it has since almost doubled. Jet aircraft, much noisier than piston-engined planes, came into regular service in 1959 and the ultimate in aircraft noise, Concorde, began scheduled flights in 1976. Large-scale objective measurement of noise in Britain began in 1948 at points in the open air. Road traffic accounted for most of the noise (84 per cent) followed by industry (7 per cent) and railways and building (4 per cent each). It is surprising that aircraft noise was apparently negligible as recently as 1960, since it is now a considerable source of noise, in all but the busiest streets, in those parts of London that lie below the flight paths to Heathrow. Subjectively, the impact of noise is easier to measure since elaborate apparatus is not needed – only a social investigator with a clipboard and pencil. Sensitivity to noise, it appears, varies widely. A survey in 1960 found that 11 per cent of a large sample of people (1400) put noise at the top of their list of complaints; a similar proportion objected to slums, dirt and smoke, and to the kind of people they met or had as neighbours; somewhat alarmingly, nearly a third of the sample had no complaints at all! Investigation among people living near Heathrow airport, the busiest in the country, found that a similar proportion (32 per cent) were undisturbed by noise measuring 103 decibels or more. This level of noise is as loud as a disco, and considerably louder than a heavy diesel-engined lorry at a distance of 25 feet. At the other extreme ten per cent were sensitive to all aircraft noise.

The coming of the jet engine in 1959 temporarily led to a trebling of complaints about noise at Heathrow, but by 1962 the volume of complaint had subsided, though it remained above the 1958 level. Perhaps a more telling indication of the strength, or rather weakness, of public feeling came from the evidence of house prices: the noise from Heathrow did not affect them at all, a result confirming studies of house prices in busy and quiet streets. Sensitivity to noise, as the Wilson committee predicted, has increased with affluence and rising expectations. In 1971 environmental health officers received some 13,000 complaints, a substantial increase over the level of discontent recorded in 1960. Since then there has been a seven-fold increase in the number of complaints. All categories of offenders – industrial/commercial, building, traffic, domestic – provoked extra complaints, but it was domestic noise, the noise of neighbours, that people complained about most as time went on: in 1971 domestic noise accounted for little more than a quarter of the complaints; by 1986–7 it was two-thirds of a very much larger number.

There had always been legal remedies for noise as for other forms of pollution, through an action for nuisance, and the courts gave judgement against a wide range of noises – the ringing of heavy bells in a house, a circus, a rifle gallery, the shuffling and neighing of horses in a stable, and the vibration caused by an electric power station. The principle that even a small addition to the level of pollution was offensive and actionable, laid down in *Crossley* v. *Lightowler* in 1867, was applied to noise nuisance in 1906. A Sheffield house-holder, the court declared, was entitled to protection from the additional noise of a steam-hammer, even of the most modern kind:

> If a substantial addition [of noise] is found as a fact in any particular case, it is no answer to say that the neighbourhood is noisy, and that the defendant's machinery is of first-class character. (Buckley, *Law of nuisance*, 1981, pp 23–4.)

However, as with river pollution, it was a far cry from one successful action against noise to the restoration of peace and quiet in general. Few of those affronted by noise were rich enough or obstinate enough to go to law. There was not even a society to protest about noise until 1959 when the Noise Abatement Society was founded. As befits its title the society has not been particularly vociferous. Because noise is largely transmitted through air the National Society for Clean Air has taken up the cause of a quieter Britain, and its work has overshadowed that of the Noise Abatement Society. Central government first showed interest in the problem of noise when in 1960 it appointed a committee of enquiry chaired by Sir Alan Wilson, FRS. The committee produced a level-headed, politic and constructive report, but by an accident of parliamentary politics its thunder had been stolen by a private Member. Mr Rupert Speir was lucky in the ballot for private Member's bills and in November 1959 his Noise Abatement Bill received its first reading. It passed easily through parliament and became law in the summer of 1960, almost unopposed. Members who spoke in the House regretted that the railways (by convention) were exempted from compliance and that aircraft noise (by acts of 1920, 1947 and 1949) was also beyond the reach of the law. Parliament could of course have repealed or amended the earlier acts – at the risk of ending civil aviation in Britain. Mr Speir well knew that a first attempt at a general law against noise should not try to do too much, and there was never any serious threat to discipline either the railways or civil aviation. The Noise Abatement Act made excessive noise a statutory nuisance, so that local authorities could prosecute on behalf of aggrieved or injured citizens. A second important feature of the noise abatement act was the idea of the noise-abatement zone, by analogy with the smokeless zones of the clean air act. Local authorities were empowered to establish noise-abatement zones, but the increasing number of complaints and general observation both suggest that the law has proved largely ineffective, especially in controlling domestic noise.

Relief from noise is more likely to come from one of two other directions: heightened awareness of the problem followed by a neighbourly or good-mannered response; or secondly from changes in technology. Of an improvement in manners there is little evidence, and there is no reason to expect it. The rising level of crime, raucous popular music, litter in the streets and the appeal to self-interest of recent public policy all point to an inconsiderate rather than a well-mannered society. With noise, as with smoke-abatement, the first effective steps towards a less polluted environment have come not from persons but from public bodies and firms, the corporations that have neither a body to be kicked nor a soul to be damned. Not only do corporations have ready access to the latest technology; they are also more vulnerable to the pressure of the law than a private person. Hi-fi equipment cannot compete with a steam-hammer as a nuisance and regulating the conduct of a few powerful noise-makers is easier than pursuing a host of petty offenders through the courts. Sometimes quite simple changes in technique reduce noise levels substantially. For example the replacement of cobbles and granite setts by wooden paving blocks did much to reduce traffic noise in late Victorian London. The coming of the motor-car replaced the clatter of horses' hooves with the much louder roar of an engine, partly offset by pneumatic rubber tyres, which were decidedly less noisy than the iron-rimmed wheels of wagon or hansom cabs. Insulation against sound has often been an effective remedy. Double-glazing of houses or schools is usually installed to conserve heat, but it does incidentally reduce the level of street or aircraft noise. Near airports and motorways double-glazing with the aid of government grants has provided some relief from noise. Effective protection from aircraft noise requires such thorough insulation that air-conditioning has to be installed; few can afford such expensive devices. It is possible to muffle noisy machines like the pneumatic drill. Some reduction of noise follows, but an already heavy piece of equipment becomes still heavier, and muffling is not widely practised. Noise consultants are a rare breed of consulting engineer, but that they exist at all is a sign that some firms seriously wish to lower the volume of noise, and are willing to spend considerable sums of money to avoid undue nuisance to the public. The latest jet engines are less noisy than their predecessors and aircraft noise per plane has fallen by about 20 decibels since the widespread introduction of jet engines thirty years ago.[8]

There was some concern about atmospheric pollution at least a hundred years before the skies began to clear over Britain. The Wilson committee of 1960 was virtually the first acknowledgment by a British government that noise was a problem worth attending to. And it is only in the past twenty years that there has been a small but regular flow of official publications about noise: circulars, technical reports and suggested codes of practice make up this fairly slender collection of material. The campaign against noise pollution stands now where the campaign against smoke and smells stood in (say) 1880. It is unlikely therefore that excessive noise will be banished from Britain in the lifetime of any of its present inhabitants.

REFERENCES

1. Council for the Protection of Rural England [CPRE], *Annual report for 1970*; T D Kendrick, *British Antiquity* (1950), pp 19 ff, 65–77, 101–8; *Leland's Itinerary in England and Wales*, ed Lucy Toulmin Smith, repr 1964, I, xxxvii–xxxviii, 148; C L Kingsford, *Survey of London by John Stow* (1908 edn), I, p 265 ff.
2. J W Mackail, *The life of William Morris* (World's Classics edn, 1950), I, pp 349–52; Society for the Protection of Ancient Buildings [SPAB], *The proposed restoration of the west front of St Alban's church by the freemasons of England* (1879), pp 2, 4, 5–9, 13; r.c. on historical

monuments (England), *Guide to St Alban's Cathedral* (1982), p 12; SPAB, *Notes on the repair of ancient buildings* (1903), pp 17–18, 47–9 and *90th anniversary* (1967), pp 1–5; H M Colvin, *History of the King's works* (1975), III, pt i; s.c. on ancient monuments bills (PP 1912–13 VI), qq 10–16, 126–34, 310–22, 708 ff; Historic Buildings Council, *Report for 1981–2* (PP 1982–3 XXIX), p 17; National Trust, *Report for 1925–6.*

3. M A Havinden, *Somerset Landscape (1981),* pp 207–10; [Gowers] Committee, *Report on houses of outstanding historic or architectural interest* (1950), pp 21–2, 72–3; Historic Buildings Council, *First report* (PP 1953–4 XV), pp 3–5, *Sixteenth* (PP 1968–9 XXXII), pp 5–9, *Twenty-ninth* [and last] (PP 1982–3 XXIX), pp 3–7.
4. E J Mishan, *The costs of economic growth* (Pelican edn 1979), p 163 ff; F Hirsch, *Social limits to growth* (Cambridge, Mass., 1976); M Logan, *Guide to the Italian pictures at Hampton Court* (Kyrle Society Pamphlets, no 2, 1894); Society for Checking the Abuses of Public Advertising [SCAPA], *A beautiful world,* no 1, November 1893; National Trust, *First report* (1896), *Statement of accounts 1896–7, Nineteenth annual report for 1913–14.*
5. National Trust, *Freehold and leasehold properties of the Trust and protected properties* (1937), *Report 1937–8, Report by the Council's advisory committee on the Trust's constitution organisation and responsibilities* [Benson Report] (1968), pp 13–19, *Report for 1987*; *The Sunday Correspondent,* 25 February 1990.
6. D Thomas, *London's green belt* (1970); National Parks Committee, *Report* (PP 1930–1 XVI); *Report of national parks committee* (PP 1946–7 XIII); Norfolk and Suffolk Broads Act, 1988.
7. D Hunter, *Diseases of occupations* (1976), pp 851–6; R Campbell, *The London tradesman* (1747, repr 1969), p 264; *Selected letters of Thomas Babington Macaulay,* ed T Pinney (Cambridge, 1982), p 248; *The Times,* 2 July 1986; *Parliamentary debates,* 4 March 1960, col 1596 ff; C Babbage, *Economy of machinery and manufactures* (2nd edn 1832), pp 57–8; J H Clapham, *Economic history of modern Britain* (Cambridge 1932), II, 15–16.
8. [Wilson] Committee on the problem of noise, *First report* (PP 1962–3 XXII); *Social Trends 1984, 1989*; E W and H G Garrett, *Law of nuisances,* (1912), pp 174–7; R A Buckley, *Law of nuisance* (1981), pp 23–4; A J Crosbie, 'Noise as an air pollutant' and J H Richardson and R W Smith, 'The industrial and local authority approach to neighbourhood noise' (NSCA, *43rd clean air conference 1976*); M J T Smith 'Aircraft-noise control prospects for the twenty-first century' (*52nd clean air conference NSCA, 1985*) pt I.

PART TWO

The Prodigal Economy – and its Reform?

W
E

CHAPTER SEVEN

Energy

Energy is unique among economic resources in that once fully used it is lost for ever. The intense heat that converts iron ore into iron and then steel can be used for that purpose once only. The metal cools, the heat is dissipated, raising the temperature of the air slightly, but not enough for useful work to be derived from it. If too much energy is applied to the blast furnace and it threatens to escape in the form of hot waste gases, an efficient firm will recover and use this surplus energy, but once again after useful work has been extracted from it the low-grade heat that remains has no economic value. For all practical purposes the energy has been lost. This would not present problems if the incoming energy from the sun could be converted into wood, coal, oil, alcohol, or natural gas as fast as existing sources of energy are being used up. But since this is not happening, the world's reserves of energy are being consumed at a rate that is unsustainable in the long run. Other resources, unlike energy, are not lost to the economic process for ever. Soil that is eroded comes to rest in river deltas if water-borne; if wind-blown there is a good chance that it will settle on land rather than at sea. In either case it is potentially available for cultivation. Scrap metals when available in large quantities (or in small quantities if they are of high value), are brought back into use. Even if dumped on a rubbish tip, they are recoverable with the

---

Opposite:
Oil and natural gas, though less abundant than coal, have taken most of the coal industry's markets since 1945. The refinery at Stanlow, Cheshire, on the Mersey was one of several built after 1945 to meet the increased demand for oil products.

application of enough energy. The sea itself is a repository of metals in extreme dilution, and is steadily enriched as further traces are washed into it or seep into it from below the seabed. In principle and given large enough supplies of cheap energy the sea could provide many if not all the metals in economic demand. Glass and rubber readily come to mind as materials that are easily reusable, because they have favourable physical or chemical properties – a low melting point, or solubility. Other materials like bricks or pottery offer little but their bulk to the recycler. Unpromising though they are, they can at least act as hardcore or landfill material; they are not quite as useless as the low-grade heat that remains after the expenditure of energy.

COAL RESERVES

Concern about Britain's reserves of energy has a long and fitful history. In Tudor England the iron industry of Sussex, it was alleged, threatened to deprive the country not only of fuel but also of timber for building and ship building. Like many other economic forecasts this one proved wide of the mark: it underestimated the recuperative power of woodlands managed on the system of coppices; it did not foresee the migration of the iron industry to other regions; and it neglected the substitution of coal for wood fuel. None of these events happened quickly, and shortage of wood for fuel remained a problem for the poor in areas remote from coalfields and communications until well into the nineteenth century. William Cobbett in 1832 noted the shortage of wood for fuel in Cornwall and Wiltshire – but thought the poor of Sussex had nothing to complain about on that score! Alarm about wood was as nothing to the periodic alarms about coal. There was a time in the eighteenth century when miners, if not geologists, believed that mineral reserves would grow again in abandoned workings. But this comforting view was never widely held, and most people who gave the matter any thought took the common-sense line that coal, like copper, tin, zinc or any other mineral, was a wasting asset, sooner or later exhaustible. On these grounds King James VI and I regretted the export of Scotch coal:

> To communicate this so necessary a commodity to strangers, from whom we can expect none the like when we should have the want thereof, might very well be imputed for a point of negligence of the commonwealth. (Nef, *Coal industry*, 1932, II, p 225.)

Negligent or not, the export of coal – English and Welsh as well as Scottish – grew, and reached enormous proportions in the years before the First World War. In what was and long had been an essentially profit-oriented individualistic society, restriction of exports was not a practical possibility. Government never ventured beyond an export tax on coal, and even the threat of export taxes was given up under the terms of the commercial treaty with France in 1860. Outsiders might worry about the long term, but the industry – owners and managers alike – responded like the mining engineer John Buddle in 1830. Giving evidence about the expected life of the Durham-Northumberland coalfield he was asked if it would not be prudent to discourage exports. He replied:

> I do not see how the property of private individuals could be interfered with. Any man possessing a mine or quarry must be left to his own discretion as to the managing of it. (S.c. on coal trade, 1830, p 317.)

Only academic geologists agreed with King James two hundred years earlier that export of coals was 'permitting foreigners to consume the vitals of our own posterity . . . I think it is our duty to spare not one ounce of coals to any person but ourselves.' The parliamentary committee was more interested in cheap coal for London. Though it took a good deal of evidence about the possible extent of coal reserves in the north-east, that was not, perhaps wisely, its main concern. The experts after all were not agreed among themselves about the extent of the reserves. On the most sanguine view and assuming a constant rate of output, supplies might last 1700 years. Others thought 400 years the limit. None of them could foresee the extent of technical progress. In 1830 few pits had been sunk to a depth of 1000 feet and the greatest feasible depth of a coalmine was supposed to exceed that limit by only a few hundred feet. It would have greatly surprised them to be told that some British pits would eventually reach depths of 4000 feet, and that gold mines on the Rand would be worked at the prodigious depth of 10,000 feet. The committee recognised that coal was a wasting asset, but suggested no economies other than sale by weight rather than volume.[1] Like rivers pollution, the destruction of coal reserves seemed to be an inevitable result of industrial development.

Anxiety about coal reserves recurred in the 1860s after Edward Hull had compiled new estimates of the extent of Britain's mineral wealth. When Sir William Armstrong gave his presidential address to the British Association for the Advancement of Science in 1863, he drew on Hull's work. Thanks to improvements in mining technology, he felt it reasonable to suppose that all coal in seams two feet thick down to a depth of 4,000 feet could be mined in due course. Sir William cautiously assumed that the demand for coal would grow by arithmetic, not geometric, progression. In other words he assumed that demand would grow by a fixed amount each year – 2.75 million tons – rather than by a fixed percentage. On this fairly conservative basis he calculated that the country's known reserves of coal would last 212 years, whereas at an unchanging rate of demand they might have lasted over 900 years. This was alarming enough, but worse was to come. In 1865 the brilliant young economist W S Jevons published *The Coal Question*, a startling book that aroused the same sort of concern among his contemporaries as Rachel Carson's *Silent Spring* (1962), and Dennis Meadows' *Limits to Growth* (1972) in more recent times. Jevons took as his starting point the recent (percentage) rate of growth of coal output, that is, he used a geometric rather than an arithmetic progression. At the 3.5 per cent growth rate that he observed, output would double in little more than 20 years and would multiply more than 30 times in a century. Such are the powers of compound interest and statistical extrapolation! Making these assumptions and using Hull's estimates of coal reserves Jevons concluded that a hundred years would exhaust Britain's coal. As a good economist he cast about for substitutes – oil, then costing thirty to sixty times as much as coal; wind and water power; hydraulic power; electricity; the tides. He dismissed them all, refused to contemplate legal restraints upon the use of coal, and concluded, lamely, by recommending reduction of the national debt. Like Malthus, he found it easier to alarm, than to prescribe for, the patient.

The Coal Question led to a debate in the House of Commons and to the appointment of a royal commission. Hussey Vivian, the Swansea copper-smelter, made a visionary speech in moving for the commission. He was confident that the difficulties of mining coal at great depths could be overcome and, what was more striking, he envisaged the substitution of oil for coal, at sea and on land. The royal commission, 'the coal commission', published a weighty volume of evidence and conclusions in 1871. It revised the estimates of workable reserves upward to 146 billion (a billion equals one thousand million or 10^9) tons, of which 56 billion were somewhat conjectural, lying in assumed extensions of known coalfields. In addition, the commission believed that a further 41 billion tons lay at greater depths than 4,000 feet, most of it not more than 6,000 feet below the surface. Some of this coal, it was believed, lay in south-east England; only the distinguished geologist Murchison dissented from this view. In principle he has been proved wrong, though the Kent coalfield developed between the wars has not contributed largely to supplies.

The commission's forecast of future output was less alarming than that of Armstrong or Jevons. It mistakenly supposed that coal exports would not grow much; and home consumption per head, it believed, would stabilise within about forty years at a level about 17 per cent higher than in 1871. Demand would therefore depend on the rate of population growth in Britain. Estimates of future population were prepared for the commission by Price Williams, a well-known statistician. Although the birth rate had scarcely begun to fall when Price Williams set to work, his estimates did not much exaggerate the growth of numbers over the next century. The actual population in 1941 was less than four per cent, and in 1981 only thirteen per cent, below his estimate. In another hundred years it will be possible to see how far numbers fall short of his forecast of one hundred million. On the basis of these assumptions and estimates the commission made forecasts of the home consumption of coal until AD 2100. Down to 1914, since they did not foresee the huge growth of exports, output turned out to be substantially higher than they had forecast. Between the wars home consumption fell below their estimates, but exports though disastrously low for the industry remained well above the levels the commission had expected. Output as a result approximated quite closely – and accidentally – to the forecast. After the Second World War exports almost disappeared, and home demand began to fall in the 1950s. By 1981 coal output was less than half what had been forecast and was destined to fall much further. The comforting conclusion that coal supplies would last 360 years seems likely to prove right – for the wrong reasons.

No sooner had the coal commission completed its work than an inflationary boom pushed prices to record heights. The industry enjoyed a brief period of exceptional profits, and miners, it was alleged on slender evidence, toasted their own good fortune in champagne. Had the high prices of 1872–73 persisted they would have given credibility to the pessimism of Armstrong and Jevons, but the boom was followed by a long period of low prices and apparent glut, and it was not until 1899 that high prices once more brought coal reserves to public attention. A new royal commission reported that proven reserves to a depth of 4,000 feet were some ten per cent larger than had been estimated thirty years before. Though pessimistic about substitutes for coal, the commission was not alarmed, expecting coal output to grow only slowly, before stabilising and eventually declining. By the time they had completed their exhaustive enquiries in 1905, coal prices had fallen back sharply from the famine levels of 1900, and any sense of urgency had departed. Practical men do not feel much alarm when known reserves can guarantee the current level of output for a period of 400 or 500 years.

Subsequent discussion of reserves has been coloured by the thought that one generation's worthless rock may be the next generation's reserves, or vice versa. New discoveries, new technology, or higher prices may put a value on previously disregarded resources; substitutes and low prices may devalue reserves that had earlier been highly prized. Estimates made at the end of the Second World War drastically reduced the figure for proved reserves from 100 billion to 40 billion tons, or two hundred years' supply at current rates of output. The Coal Board accepted these figures without alarm though it feared that the scope for developing areas of coal as yet untouched was strictly limited. The Board's aplomb is somewhat surprising since coal supplied almost 90 per cent of the country's energy in 1950, and it was still not thought that oil, or hydro-electric, or nuclear power would make serious inroads into its markets. And nobody dreamed in 1950 that natural gas from beneath the North Sea would for a time at least be coal's most dangerous competitor.

SUBSTITUTES FOR COAL

Natural gas or methane was discovered at Hawkhurst in Sussex as early as 1836. Its presence in mines producing bituminous coal rich in volatile gases had of course long been known, but there it represented a danger rather than a commercial opportunity. Other small deposits of natural gas were found near King's Lynn and at Bassingbourne in Cambridgeshire. Early this century American businessmen established a company, Natural Gas-Fields of England, to exploit these resources. They found methane at Heathfield in Sussex at moderate pressures and at shallow depths, but not in quantities large enough to sustain a profitable business.[2] Large gas fields were found in Holland and off the Dutch coast after the Second World War. These discoveries stimulated the search for natural gas in the British sector of the North Sea. In the 1950s the gas industry started to enrich town gas with oil and in 1964 with natural gas imported from North Africa. Oil companies had already begun to prospect for gas in the North Sea, and in 1965 British Petroleum made the first substantial discovery. Commercial production began in 1967, and within 20 years natural gas was supplying a quarter of Britain's energy. When compared with the known reserves of coal, Britain's natural gas resources seem rather slender, and on present knowledge are likely to be exhausted in about 40 years. However, large tracts of the North Sea have yet to be explored, and mainland Britain may well prove to be richer in gas than has so far been demonstrated. The oil companies that do the prospecting have to pay licence fees to the British government and it cannot be in their interest to underestimate the difficulties of exploration, or to invest capital in discovering oil or gas fields that may not be needed for many years. It is useless, in other words, to expect the whole extent of resources to become apparent at such an early stage in the life of the North Sea fields. That they will be exhausted before the coalfields seems likely, but nobody can confidently pronounce when the last cubic metre of gas will be piped ashore.

Similar considerations apply to North Sea oil. Seismic exploration began in the early 1960s and serious drilling in 1964, after the award of the first round of licences. Commercial quantities of oil were struck in 1969, and at enormous expense oil began to come ashore in 1975. By 1980 Britain produced enough oil for her own needs and then rapidly became a large net exporter, with dramatic benefit to the balance of payments. Large imports of crude oil continued because oil is not a homogeneous product, and North Sea oil did not provide all the grades needed in Britain and its refineries. The oil under the North Sea is being consumed in larger quantities than the gas. Not only does oil provide a third of Britain's energy (as against a quarter from natural gas); exports represent as high a proportion of output as coal exports did in the prosperous years before 1914. On present information, output of North Sea oil is unlikely to grow much more, and reserves will last no longer than those of natural gas. These findings almost certainly err on the side of pessimism, but there seems little doubt that coal will eventually regain its position as the main fossil fuel consumed in Britain.

Set against the 1,500 years since the Roman legions left Britain, or the 10,000 years since man first became a city-dweller, the period within which fossil fuels, including coal, are likely to be exhausted is short indeed. World reserves are enormous by comparison with those of Britain, but their expected life does not suggest that once Britain's fossil fuels are exhausted an indefinite supply can be imported instead. Alternative technologies may come to the rescue. Hydro-electric power is unlikely to make a large contribution to Britain's energy needs. Windmills and wave-power have yet to prove their worth. Nuclear fusion and the direct use of solar energy are both distant and uncertain prospects. Only nuclear (fission) power is an established new technology that could, at great expense and some long-term risk, replace fossil fuel, as it largely has in France and Belgium. Those who take long views find future energy prospects daunting. If they formed a large section of opinion their views would presumably affect economic behaviour. The evidence suggests, on the contrary, that hardly anybody allows distant prospects to sway him. A house with a 999-year lease will sell for no more than an identical property with only 50 years of the lease remaining. An economy like Britain's with perhaps 50 or 100 years' supply of oil and natural gas functions as if reserves were infinite, and it is only in exceptional years of inflationary boom that the market economy betrays any unease.

ENERGY PRICES

If prospective but distant shortage aroused concern, it ought to be detectable in prices and perhaps in royalty payments. So far as royalties are concerned it is hard to detect any such trend. The major influence on the level of royalties appears to be the degree of difficulty in winning coal. When mines were shallow the medieval Bishops of Durham could exact as much as one-third of the pithead price for the original and destructible powers of the soil.* In the

* David Ricardo (*On the Principles of Political Economy and Taxation*, ed Sraffa, p 68) recognised that royalties represented a payment for the consumption of capital, and were therefore not to be confused with rent. Later (pp 85, 329–32) he forgot this important distinction, which was also

later seventeenth century a few landlords could still demand as much as a quarter of the pithead price in royalties, though an eighth was more common. In the eighteenth century, practice continued to vary: a fifth was a high proportion, an eighth more frequent; and over a small sample of collieries (26 at various dates from 1717 to 1830) royalties averaged 13.5 per cent of the pithead price of coal. The scattered information available for the nineteenth century suggests that royalties did not often amount to as much as an eighth of the pithead price. In the 1830s one-tenth or one-fifteenth was usual, and similar proportions are reported for the 1870s and 1890s. The Bishops of Durham, who had been able to claim royalties of one-third 400 years earlier, received only an eleventh or a twelfth on a sliding scale related to the selling price of coal. A more systematic but small-scale study of royalties in fifteen collieries shows that they were a greater burden on mine- owners in the later 1870s and about 1900 than in the years before the First World War. Measured against costs (presumably lower than prices in most years) royalties were at their peak in 1900 (ten per cent of costs) and not much lower in 1874–8 (nine per cent of costs); from 1902–13 they averaged less than five per cent of costs. Between the wars they fell even lower and were nationalised in 1938. None of this suggests any long-term fear that supplies of coal might be exhausted. Falling royalties indicate a greater difficulty in winning coal: that in turn reduces the relative value of the virgin seam, and denies the landlord large windfall profits from a wasting asset.

The course of prices confirms the impression left by a study of mining royalties. If energy shortages were becoming acute, the price of coal and of other fuels ought to have risen relative to other commodities. There is evidence in the sixteenth century for an exceptional rise in the price of timber and wood fuel. For example, firewood increased in price elevenfold between 1500 and 1640, a far

---

(continued from previous page)

unknown to J S Mill. The mining engineers who gave evidence in 1857 to the s.c. on the rating of mines (PP 1857 XI) took it for granted that royalties represented payment for the consumption of capital. The Cambridge economist, W R Sorley, followed their views when laying down the modern theory of royalties in 1889 (Journal of the [Royal] Statistical Society). His reasoning convinced the royal commission on mining royalties, 1894, but not the Inland Revenue. Mining royalties continued to be regarded as income for taxation purposes, rather than as a realisation of capital. In 1970 it was conceded that life annuities contained an element of capital repayment, but the same concession was not made to owners of royalties.

higher rate of increase than for prices in general. But for coal prices no such pattern appears. For the seventeenth and eighteenth centuries the longest series are contract prices paid by Westminster School and Greenwich Hospital. Contract prices are notoriously sticky and suspect; these particular prices include varying levels of tax and the costs of transport from Newcastle by sea, from ship to shore and from shore to the institutions' coal cellars. The largest of the price movements, apart from those due to taxes, were caused by the Dutch and French Wars. There was no tendency for the price of coal to rise faster than other prices. For the past 200 years more trustworthy price series are available, often untaxed and embodying fewer transport costs. They do not show any long-run tendency to increase faster than other prices.

In the nineteenth century and even in years of exceptionally high prices – 1800, 1814, 1873 – coal prices kept below the general advance. And in periods of deflation, coal fell at least as fast as prices in general. The year 1900, a time of widespread alarm, was one of the few when coal prices rose faster than the average. In the period from 1930 to 1960, cartels and government policy distorted coal prices, and until the past ten years all the nationalised industries were subject to short-run political interference in their pricing policy. The rapid increase in oil consumption after the Second World War seriously affected the market for coal and like political interference depressed prices and the profits of the Coal Board. Since 1973 the oil industry's cartel, OPEC, has twice succeeded – in 1973 and 1979 – in sharply increasing oil prices. In 1973 it was noticeable that fuel prices were already advancing of their own accord before OPEC gave them a much sharper upward push. The second increase was short-lived since OPEC was trying to maintain prices at a time of falling demand and of increasing supply from non-OPEC sources. In 1990 the threat of war in the Middle East raised oil prices once more, but not to the 1980 levels. When allowance is made for general inflation, it does not appear that oil prices have diverged much from general trends.* When recourse is had to the retail price index, which is a satisfactory

* The real price of petrol, which is a manufactured product far removed from crude oil and which embodies a high level of tax, was lower in 1990 than thirty years before. The highest real prices of petrol were recorded during and after the Suez crisis (1956), in 1975, and in the early 1980s (*The Independent*, 12 September 1990).

account of prices since 1947 though not before, it appears that fuel and light rose somewhat less than other prices between 1947 and 1974, and rather faster after 1974.[3] As a pointer to energy scarcity retail prices are less satisfactory than wholesale prices, but for what they are worth they give little support to theories of fuel shortage.

ECONOMY IN THE WINNING AND USE OF COAL

One of the difficulties of estimating reserves of coal, oil, or gas is that improving techniques may allow a larger proportion of the raw material to be recovered. Some of the early methods of winning coal were notoriously wasteful. The 'pillar and stall' technique was favoured in South Wales in the mid-nineteenth century, and on the north-east coast until well into the twentieth century. As its name implies, pillars of coal were left in the pit to support the roof (and sometimes as at Wigan to prevent subsidence in the town overhead). It was often feasible to rob the pillars of some of their coal as the men worked their way back from the limits of the mine to the shaft bottom, but commonly ten, twenty, or even thirty per cent of the coal was left underground. An alternative technique – longwall mining – was practised in Shropshire as early as the seventeenth century. Under this system the hewers advanced through the coal seam on a broad front, filling the empty space left by their excavation with the rubbish of the mine – stone from the roof, floor, or sides. The longwall method was not necessarily appropriate in all geological conditions – it needed an adequate supply of rubbish for back-filling, and it worked best where there was little danger of rock movement. Where it was appropriate, however, it allowed nearly all the coal to be removed from a mine, and knowledgeable managers had every incentive to adopt the system wherever possible. News of the improved method travelled slowly. Shropshire miners were imported into Lancashire in the 1760s to work the Duke of Bridgewater's mines. Ninety years later the system reached South Wales, and within twenty years had largely displaced pillar and stall working there.

There were other sources of waste in and about coalmines. Small coal was left underground if it was only weakly caking, as with Welsh steam coal. Strongly caking coals had more value above ground because they could be used to make coke. When coal-washing became common in the second half of the nineteenth century a great deal of coal dust was washed into rivers until colliery owners under pressure and out of self-interest found ways to prevent the loss. There can be no doubt that periods of high prices, as in 1873, encouraged owners to recover as much coal as possible. When prices fell again the good habits would not necessarily be abandoned. Some of the coal mined was consumed by colliery steam engines. Since coal at the pithead was relatively cheap there was little reason to save coal by investing in superior steam engines. Nevertheless, competent observers thought that the efficiency of colliery engines in Lancashire had improved by 50 per cent in the last thirty years of the nineteenth century, and that no more than a fortieth of the coal mined was used for steam raising at the colliery itself. Other coalfields remained technically backward in their methods of steam-raising, and wasted huge quantities of coal. By 1947 electrification had displaced many colliery engines. But in north Staffordshire at nationalisation 22 collieries still had 169 boilers, mostly hand-fired, at work. The coalfield produced about seven million tons of coal, of which no less than half a million tons served the colliery engines.

A colliery engine-house would never have been the most promising place in which to look for fuel efficiency. Small coals, which were particularly cheap, were readily available. They tended to give off clouds of smoke, but since most pits were sunk away from large towns few troubled about the smoky air. The keenest exponents of fuel efficiency in Victorian Britain were firms at a distance from coalmines where transport costs made coal an expensive commodity. It was no accident that Cornwall and London boasted some of the most economical steam engines. The amount of energy or useful work in a pound of coal can be calculated with accuracy in a laboratory. The earliest Newcomen engine captured about 0.5 per cent of this useful work; a modern turbine in an electricity power station has a thermal efficiency of about 40 per cent, that is, it does eighty times as much work for a given quantity of fuel as the Newcomen engine. Much higher efficiencies are possible with systems of district heating (combined heat and power, or CHP), but this way of capturing waste heat from power stations is not much practised in Britain. Higher efficiency is not pure gain. Engineers save fuel only by devising more elaborate machinery, in other words by substituting capital for raw materials and labour. The capital embodies fuel and other minerals, as well as labour, but if fuel efficiency is pursued with due regard for costs, the saving in fuel will exceed in the long run the cost of the extra capital required.*

Except at a few collieries, the low-pressure steam engine devised by Newcomen, improved by Smeaton and transformed by Watt, was everywhere replaced as the nineteenth century wore on by more efficient and more powerful high-pressure engines. This is not the place for an elaborate account of the steps which have led from the comparatively simple Newcomen engine to the powerful turbines in a modern power station. But anyone who ventures to chronicle waste and the search for efficiency must at least sketch the technical history of the steam engine, a major source of power in the modern world. It is convenient to consider the steam engine's efficiency under three headings: the fire that converts water into steam; the boiler that contains the water and steam; and the engine that converts the power of steam into work. Something has already been said about the fire in discussing air pollution, but avoidance of smoke is not necessarily proof of efficient combustion, and other steps need to be taken to secure maximum heat from the fuel. Smoke at its worst accounted for only a small proportion of the fuel consumed – about one per cent. A smaller quantity was deposited as soot on boiler plates or in the chimney stack and had to be removed at regular intervals for efficient combustion to proceed.

A small but useful gain in efficiency occurred when after the Second World War smoke was virtually eliminated from factory operations. Larger gains were secured through the complete combustion of volatile gases and by preventing the formation of

---

* The economist regards a more efficient steam engine as a substitution of capital for land (coal) and labour (the work of coal-miners). He may add that the improved steam engine also embodies a fourth factor of production, enterprise. For purposes of exposition these are convenient terms, but in reality land and labour are the only factors of production: enterprise is a special kind of labour; and capital is embodied land and labour. Capital in the sense of money means access to, or command over, land and labour. When economists dismiss fears of resource shortages and suggest that any difficulties can be surmounted by substituting capital for land, e.g. a hydro-electric for a coal-fired power station, they are forgetting that capital is not an autonomous factor of production. It is true, of course, that in the long run the hydro-electric plant will consume less 'land' than one fired by coal, and that the world's resources will last that much longer as a result. But all capital investment makes some inroad into finite resources of 'land'. An alternative view that land is a part of capital is discussed on pp 247–50.

cinders. It was easy to overdo the attempt at complete combustion: too high a temperature in the furnace converted carbon dioxide back into carbon monoxide with consequent loss of energy, especially if the gas then escaped up the flue. It has required nice adjustment of furnace conditions to eliminate this loss. The use of powdered fuel, introduced into power stations and cement works before 1939, has got rid of the problem of cinders. Five or even ten per cent of the coal burnt could be lost unless steps were taken to recover cinders by hand-picking or with the aid of a special fork. The change to electric power therefore went far to get rid of this important energy loss, even after allowing for the cost of coal-crushing. In the latest type of furnace with fluidised-bed combustion the distinction between fire and boiler is hard to maintain. In order to gain a high rate of heat transfer from the fire to the boiler plates or tubes, the boiler is immersed in the fire bed. This consists of a deep layer of heated ash or sand into which air and powdered coal are injected from below. By keeping the temperature below 1,000°C clinkering can be avoided, and efficient combustion and heat transfer are assured. Work on fluidised-bed combustion began in France in the 1950s, was taken up by the National Coal Board in the 1970s and has had some useful applications. Its success does not appear, however, to be either rapid or certain.[4]

Fluidised-bed combustion is a late example of attempts to transfer heat more efficiently from the fire to the boiler. Richard Trevithick (1771–1833) was the first to see that a fire external to the boiler was less than economical. His Cornish boiler had the fire grate and flue encased in a cylinder inside the larger cylinder that was the boiler proper. In this arrangement, a larger heating surface was presented to the flames and hot gases generated by the fire, increasing the quantity of steam raised per pound of coal. In 1844 William Fairbairn went one better with the Lancashire boiler, which had twin fires, both encased in the boiler. Both the Cornish and the Lancashire engines, with appropriate modifications such as mechanical stokers, economisers and superheaters, held their own as steam-raisers in industry until after the middle of the twentieth century. The boilers could be as much as thirty feet long and up to six or even ten feet in diameter. Such a large body of water and steam in one undivided compartment called for strict maintenance and intelligent management, all the more necessary until accurate pressure gauges were invented in the 1850s. Inevitably, poor design, bad workmanship and inattentive supervision led to some disastrous accidents, for a malfunctioning Lancashire boiler had the force of a powerful explosive charge. In 1855 a group of Lancashire industrialists and engineers founded the Manchester Steam Users' Association. Through its inspectors the association measured the efficiency of boilers, inspected them (at first externally and from 1864 internally as well), advised on smoke-free stoking, and guaranteed members' boilers against explosion. In its first ten years, before boilers were internally examined, eight explosions took place at members' premises, but in the next 40 years there was only one such explosion although the 2,000 members worked nearly 9,000 boilers. There was no statutory regulation of boilers until 1882, and as many as 90 men could be killed in a year among non-member firms. The usual number of such deaths was in the range 50–80. After the Boiler Explosions Act of 1882 deaths were usually below 20 and never exceeded 40 in a year.

This welcome reduction in the number of accidents and deaths occurred despite higher boiler pressures. Higher pressure implied higher temperatures and severe working conditions for boiler plates and other components. The very high pressures in modern turbines demanded special steels, and until they were available the full economies of high-pressure working could not be achieved. Once the metallurgical difficulties had been overcome, fuel savings more than offset the extra capital costs. The savings arose from a peculiar property of steam: under high pressure water boils at a higher temperature, but less extra energy or latent heat is needed to turn it into steam. At normal pressure it takes 970 British thermal units (BTU) to convert a pound of water at boiling point into steam; at a pressure of five atmospheres only 895 BTU, and above about 200 atmospheres' pressure (3206.2 pounds per square inch [p.s.i.]) no latent heat is needed at all! The extra capital cost of working at very high pressures limits the ambition of engineers, and the latest power stations operate at pressures considerably below the theoretical optimum if fuel saving were the only consideration. Nevertheless, the balance between capital costs and fuel costs has been tempting engineers, from Trevithick onward, to strive for ever higher pressures. Exact figures do not exist, but it is possible to give a broad indication of pressures commonly attained. In 1854 the Lilleshall Company of Shifnal, Shropshire, one of the largest coal and ironworks in the country, had a hundred steam engines, whose power ranged from six to one hundred hp; the pressures at which they worked ranged from five to 35 lb psi. In 1869 typical pressures in a survey of over 2000 boilers were from 31 to 60 lb psi.; and hardly any of the boilers were working above 75 lb. By the end of the nineteenth century new engines installed at collieries were working at 160 lb psi.

The danger and expense of high pressures as well as the urge to transfer heat efficiently led to the development of the water-tube boiler. Instead of one large undivided compartment, the water-tube boiler had a large number of small tubes: if one burst the danger was small; and for a given volume of water a larger surface area was presented to the heat of the fire. Difficulties of manufacture and perhaps the conservatism of British industrialists prevented the rapid spread of the new boiler. The American firm of Babcock & Wilcox Ltd patented the first commercially successful version in 1867 without much effect in Britain. Only at the end of the nineteenth century were electricity companies and the Royal Navy coming to specify the water-tube boiler, especially for use with turbines; for general industrial purposes the Cornish and Lancashire boilers continued to hold their own. With proper maintenance a boiler might last as long as 40 years. Its running costs were small compared to a firm's total costs, and the average firm might not trouble itself with small economies of fuel until faced with the need to replace an old boiler. New firms would naturally order the latest and most efficient plant, but it took a long time to raise average standards in this way.

The most rapid advance in boiler technology occurred at power stations. Early in this century boilers were being designed to operate at pressures of 180 lb psi. By the 1940s 650 lb was 'relatively common'; in the 1950s pressures reached 1,500 lb at Castle Donnington, and in the late 1960s, 2,300 lb became standard practice. Except at Drakelow C the industry has rested content with these considerable pressures. Consumption of fuel per unit (kwh) of power generated has fallen steadily with higher pressures, larger generating sets, and fewer larger stations. At the beginning of this century 8 lb of coal were needed to produce a unit of electricity, in 1921 3.4 lb, and in 1938 only 1.4 lb. Thermal efficiency of generation in 1938 was rated at about 21 per cent. It has steadily increased ever since, reaching 27 per cent in 1966 and 35 per cent in 1987. The huge coal-fired station at Drax achieves an efficiency of 37 per cent. Over the same period transmission losses have fallen. The early power stations offered supplies only within a narrow radius, yet still lost a sixth or even a fifth of their power in transmission. The establishment of the national grid in the 1920s greatly increased the scope for losses, but high voltages offset this tendency and in 1950 losses were no higher than a sixth (c.15 per cent). The extremely high voltages at which power is now transmitted have sharply reduced losses to no more than 2.5 per cent.[5] Roughly speaking twelve times as much power can now be delivered for a given quantity of coal as was possible with the techniques of ninety years ago. It is not surprising, therefore, after all allowances are made for the costs of large power stations and the national grid, that electricity has acquired a substantial share of the market for heat and power, and supplies practically all of the lighting in Britain.

It is not often possible to achieve such spectacular savings in fuel even over a long period of time. Nor is it possible in a short chapter to indicate all the other more modest ways in which alert management has secured economies. What follows is a small sample of improvements that have in total contributed to extracting a greater quantity of useful work from a given quantity of fuel. One obvious but often neglected way to save fuel was through lagging and insulation. To lag steam pipes and protect a steam engine and its boiler from wind and weather might seem obvious precautions. Yet cheap coal allowed firms to prosper without taking these elementary steps. The Coal Commission of 1867–71 received evidence from several engineers about wasteful practices. James Watt had devised a system of steam-jackets for boiler cylinders, but the idea was still not widely adopted in 1870, and only one of the Lancashire boiler-makers at that time offered steam-jackets with its boilers. The lagging of pipes, like the training of stokers, was a perennial subject of discourse; if engineers never tired of it the suspicion arises that firms were neglecting a simple economy. Their neglect may not at first have been unreasonable. The early lagging materials included wood, cork and, most effectively, loose wool, all of them highly combustible. By the twentieth century inorganic materials were available, safer but less efficient at conserving heat: the best of these was 'mineral wool' or aerated slag from the blast furnace; somewhat less efficient was asbestos. Later, aluminium foil came into use. For the very high temperatures found in blast furnaces, kilns, and coke ovens, refractory bricks had always been necessary. They not only prevented heat loss but also protected the metal plates from corrosion. Insulating bricks – light and porous – were no substitute for the protective role of refractories, but supplemented their heat-saving properties. The Ministry of Fuel and Power was still advocating their use in 1944.

Heat loss in the Englishman's home had long exercised

engineers. What little heat was left after eight- or nine-tenths had gone up the chimney rapidly escaped through the windows, walls and roof. The value of double-glazing in adding to comfort in the home or office was demonstrated as early as 1857. Experiments conducted for the General Board of Health showed that double-glazing with the panes set five inches apart raised the temperature of rooms by as much as 9°F. Ministers and senior civil servants might well have acted on this discovery for their own comfort. Measurement in the board room of the GBOH showed a sorry state of affairs. The room was fairly large with high ceilings (33' × 22' × 14'), its five windows faced north and west, and heat was provided from one small fireplace. When heated by an efficient stove a comfortable temperature was secured only within about six feet of the fire. At a greater distance the temperature was raised only 15°F above the temperature outdoors. The observations were made in the winter when open-air temperatures ranged from 30°F to 45°F. In other words, in the coldest weather most of the boardroom of this government department was heated to about 45°F. There is no reason to think that other government departments, or great country houses, fared any better. More than 130 years later, full double-glazing remains unusual: it can make a substantial contribution to warmth and quiet, but the cost is high and most people stop short at central heating. Heat loss through cavity walls occurs on an even larger scale: three times as much heat disappears through cavity walls as through windows. More would have escaped before cavity walls were introduced, about the year 1870. Like double-glazing, cavity wall insulation has not proved very popular; though it is comparatively cheap, the fear of damp penetration has discouraged many. The easiest and cheapest ways to avoid heat loss in the home are roof insulation and lagging the hot-water tank. Neither job calls for skill or paid labour; the materials cost little; and many householders have found it worthwhile to spend time and money in this way. Advertisers and official encouragement have proved thoroughly persuasive. The advantages are plain to see, there are no serious drawbacks, and there is no need for elaborate calculations about pay-back time and rates of return. Except for the lagging of the hot-water tank, none of these possible economies was widely known, still less practised, thirty years ago.[6]

Householders had few ways to save heat: lagging, insulation and the use of efficient stoves or grates exhausted the list. For industrialists a huge number of variations on these themes was possible. For brevity's sake only a few that were either widely applicable or made large savings will be mentioned here. In many industries evaporation, usually of water, but sometimes of other liquids, is a necessary process, and at normal pressures uses a considerable amount of heat. As every schoolboy knows, pressure decreases with height, which is why it is difficult to cook an egg on a mountain top – the water boils (evaporates) at too low a temperature. At high vacuum (0.5 lb psi) water will boil at 80°F (27°C), and vessels strong enough to resist atmospheric pressure can evaporate water without using much heat. There was no technical difficulty in building such vessels since steam boilers were successfully resisting pressures of two or three atmospheres in the 1860s and of ten atmospheres by 1900. Evaporation under reduced pressure, but not high vacuum, was tried in sugar refining in 1812 and in the distillation of whisky in 1817, but systematic use of the technique had to wait for many years. Its use in saltworks would not only have saved fuel, but reduced the emission of hydrochloric acid gas. As late as 1876 the saltworks of Cheshire were still evaporating water from a thousand open pans without any regard for the help that science could offer. The trade had grown enormously from small beginnings in the eighteenth century, and by 1876 1.75 million tons of salt were obtained from brine after evaporating three times that weight of water. In 1906, when the trade had slightly declined, only two of the 53 Cheshire saltworks employed vacuum evaporation. By 1962, 70 open pans survived making about an eighth (150,000 tons) of the white salt produced; the rest came from vacuum plant. By then vacuum evaporation had become less important since four million tons of brine containing about a million tons of salt were used direct in the chemical industry. By the First World War the technique so belatedly adopted in saltworks was widespread elsewhere, in the alkali trade, the manufacture of condensed milk, textile processing, the production of glycerine and the fractional distillation of oil.[7]

The iron industry was for long one of the most extravagant consumers of energy. The uncapped blast furnace, it was estimated, allowed 80 per cent of the heat generated to escape, much of it in the form of carbon monoxide and light hydrocarbons. It was possible, even without capping the furnace, to draw off some of the hot gases for heating the blast. From the middle of the nineteenth century the more venturesome ironmasters began to experiment with capping, and the practice slowly caught on. Most Scottish furnaces remained uncapped in 1882, but by 1900 the open furnace was a rarity there; Derbyshire and South Yorkshire were its last strongholds in England. J S Jeans, a well known authority on the iron and steel industry, estimated at the beginning of this century that more than two million tons of coal were still being wasted every year through uneconomical practices at the blast furnace, despite many improvements made in the previous century.

Neilson's hot blast of 1828, for example, reduced (it was said) the amount of coal required to make a ton of pig iron from eight to only three tons. Later and more careful measurement showed that the saving was much less – what was gained in the furnace was partly lost in heating the blast. In 1831 furnaces at Dowlais using a cold blast needed two and a half tons of coal to produce a ton of pig iron. In 1837 the same firm needed from two tons one hundredweight to two tons four hundredweight using the hot blast. This seems to be a considerable saving, but the best coal – 'stone coal' or anthracite – was being used; and it is not known whether the same kind of coal was used in 1831. Coal consumption also depended on the richness of the ores to be smelted. It is not likely that there was much variation in quality between 1831 and 1837 at Dowlais, but any comparison between different regions or over long periods of time is likely to be affected by variations in the ore. The evidence suggests that economies in fuel consumption were achieved only slowly. By 1870 the Dowlais furnaces averaged 38 cwt of coal per ton of pig iron. Other regions using coal of a lower calorific value needed rather more – about two and a half tons. At the end of the nineteenth century the average had fallen to about two tons. By 1956 one ton of coke, equivalent to 30 or 35 cwt of coal, was all that was needed. And by 1985 consumption of coal or other fuel (coke, oil) amounted to less than a ton of 'coal-equivalent' fuel.

Fuel economies in the iron and steel industry were of the greatest importance, for the industry was a major consumer of coal: in the early 1870s nearly a third of the country's coal output went into the making of iron and steel; at the end of the nineteenth century the industry still took a sixth of coal output; and in recent years it has remained, after electricity, coal's second largest customer. Other economies were, of course, open to ironmasters, some eagerly taken up, others more reluctantly. The reception given to Siemens's open hearth furnace, a rival to Bessemer's converter, was on the whole favourable. Costly to install, it saved up to half the fuel normally used for puddling iron, or making steel. After a slow start British firms had by 1870 taken out half the licences granted by Siemens. Since Britain then produced about half the iron and steel then made in the industrialised world, there seems to have been no lack of enterprise on that score at least.

Coke ovens, like blast furnaces, were long wastefully managed. Gas companies naturally took care to collect the gas given off when coal was coked because their principal business was gas-lighting, and coke, tar and ammonia were for them (sometimes troublesome) byproducts. For collieries and ironworks, on the other hand, hard or metallurgical coke was the main product, and it was not until the beehive oven gave way to more economical designs that coal gas was collected and put to good use. The Belgian mining engineer Carvès designed a new coke oven that allowed for the recovery of byproducts, and installed it at Bessèges near Courtrai in the 1860s. He found few imitators in England for many years. The benefits of his oven were small in South Wales where much of the coal had a low volatile content, in other words produced less gas, tar or ammonia than would be given off by bituminous coal. Elsewhere the beehive oven gave way more readily but it was still predominant at the end of the nineteenth century. As it slowly disappeared in this century, steelworks acquired a cheap supply of heat hitherto wasted, and colliery coke ovens had large surpluses of gas for sale to gas companies and municipal gasworks. This trade grew enormously – from 1.3 billion cu ft of gas in 1921 to 54 billion cu ft in 1947. In that year coke ovens provided one-eighth of the gas sold by gas companies and municipal gasworks. Nationalisation of the gas industry led to further economies. For example, a few hundred yards of connecting gas main allowed Sheffield to be supplied with surplus gas from steelworks in Rotherham.[8]

THE DEMAND FOR ENERGY

It would be easy but unprofitable to multiply examples of economies in the use of energy. Undoubtedly the search for efficiency in energy use will continue and be praised, not least by conservationists. However, it is a regrettable fact that efficiency is never so complete as to lessen consumption. Economists from Jevons onward have noted with perverse satisfaction that economy cheapens, that cheapness extends the market, and that measures of conservation or economy therefore increase, or at least do not diminish, the consumption of energy. What would have happened without conservation measures is less clear. The consumption of energy might have increased still faster, though this is unlikely since without increased efficiency real incomes cannot rise. It is rather more probable that less energy would have been called for, and it is certain that considerably less satisfaction or real value would have been obtained from it.

The growth of demand for energy in the UK bears out these contentions, even though precise figures for energy consumption per head are unobtainable. When the whole of Ireland was part of the UK, peat added somewhat to energy supplies. At all times in the past 150 years kindling wood and logs have to a small and decreasing extent supplemented domestic coal. In industry water power and windmills at one time made a useful contribution to the supply of energy. More important now is the contribution of oil. In the nineteenth century the major sources of oil for lighting – then its main energy use – were tallow and blubber. Oil from Scottish shale supplied some paraffin and candle wax, mostly between 1850 and the First World War. James Young, a Scots chemist, pioneered the industry, establishing works at Bathgate near Edinburgh. The process was cumbersome since each ton of shale yielded at best 30 gallons of oil and about 40 lb of ammonia, leaving a large quantity of spoil for which even brickmakers had no use. In the absence of large supplies of cheap mineral oil – the oil industries of the USA, Russia, Romania and the East Indies being as yet in their infancy – shale oil production grew until the First World War, but has since dwindled into insignificance. The oil-fired ship, the motor-car, the bus and lorry became important only with plentiful supplies of oil from wells, that is, after the First World War. H S Jevons, a more conventional economist than his father, published a long study of the coal trade in 1915. In agreement with the current orthodoxy, he dismissed the possibility of oil as a substitute for coal. That was by no means an unreasonable view at the time, for as late as 1950 oil had taken only ten per cent of the market for energy in Britain. Cheap oil from the Middle East and the establishment of large refineries in Britain then rapidly transformed the situation. By 1970 oil had overtaken coal as a source of energy. And by 1987 natural gas had also become a dangerous competitor, not far behind coal. Nuclear power, dogged by technical difficulties and distrusted by large sections of the public, lagged far behind, generating less than a sixth of the electricity and holding much less than a tenth of the total market for energy. Between them these rivals have easily offset the recent fall in home demand for coal.

Though the coal industry had been in retreat since 1913 home demand for coal went on rising until after the Second World War (see Chapter two). Per capita consumption rose slowly, increasing by a third between 1871 and 1951, most of the increase taking place before 1900. In Table 7.1 the figures for 1951 include oil, and the figures for 1975 and 1987 also allow for natural gas and for nuclear and hydro-electric power. They therefore represent coal-equivalents rather than simply coal. The figures for 1871–1925 are of coal consumption only, ignoring peat and wood, and wind and water power. Strictly speaking, consumption per head at those dates is underestimated, but only to a trifling extent.

Table 7.1 UK energy consumption 1871–1987

	Energy consumption (million tons of coal)	*Population (millions)*	*Consumption per head (tons)*
1871	105	31.5	3.3
1900	181	40.1	4.5
1925	192	45.1	4.3
1951	249*	50.2	4.9*
1975	331*	55.9	5.9*
1987	353*	56.9	6.2*

* Coal equivalent

To a conservationist Table 7.1 makes mournful reading. In little more than a hundred years the demand for energy per head of population has come close to doubling. In the first quarter of this century demand fell but only by a little. The relative increase in the price of energy since 1973 and the subsequent severe depression in manufacturing industry have slowed the rate of increase without eliminating it. If living standards continue to rise in Britain, the demand for energy will almost certainly rise too, and the prodigal economy will hurry on its way. Today's consumer gets more satisfaction from every unit of energy than consumers in the past – but that does not stop him asking for more.[9]

REFERENCES

1. N Georgescu-Roegen, *Energy and economic myths* (New York 1976); R H Tawney and Eileen Power, *Tudor economic documents* (repr 1953) I, 231–8; W Cobbett, *Rural rides* (Everyman edn 1948), 158–9; J U Nef,

The Rise of the British coal industry (1932), I, 270 n, II, 225; s.c. on the coal trade (PP 1830 VIII), pp 13, 244, 316–17, 396.

2. British Association for the Advancement of Science, *Report of the 23rd meeting* (1864), pp lii–liv; W S Jevons, *The coal question* (1865), pp 117–45, 209, 213, 339; *Parliamentary debates*, 12 June 1866, cols 241–75; Coal Commission (PP 1871 XVIII), *Report*, pp ix–xvii; r.c. on coal supplies, *Final report* (PP 1905 XVI), pp 2–6, 15–17; Political and Economic Planning, *British fuel and power industries* (1947), pp 3, 59–63; NCB, *Plan for coal* (1950), pp 20–3, 25, 41–50; r.c. on coal supplies, *Evidence* (PP 1904 XXIII), qq 11264 ff, 11294, 11337.
3. Nef, *Coal industry*, I, pp 139, 326–7; M W Flinn, *History of the British coal industry* (1984), II, pp 43–4, 292–3; R A Church, *History of the British coal industry* (1986), III, pp 507–8; s.c. on . . . coal (PP 1873 X), qq 461, 2544, 2799, 3195–7, 3543–5; r.c. on mining royalties (PP 1890 XXXVI), qq 12–27; J U Nef, 'Prices and industrial capitalism', repr in *Essays in economic history* ed E M Carus-Wilson, I, p 130; B R Mitchell and P Deane, *Abstract of British historical statistics* (Cambridge, 1962), pp 479–84; *AAS* v.d.
4. T S Ashton and J Sykes, *The coal industry of the eighteenth century* (Manchester, 1929), pp 16–32; J H Morris and L J Williams, *The South Wales coal industry 1841–1875* (Cardiff, 1958), pp 58–62; Coal Commission, *Report* (PP 1871 XVIII), ix; r.c. on coal supplies (PP 1905 XVI), pt x, qq 18721–7; s.c. on . . . coal (PP 1873 X), qq 993–4, 1169–71, 1437, 2771, 3110 ff, 3266–70; r.c. on coal supplies (PP 1904 XXIII, ii), qu 14710; NSCA, *Clean air conference 1969*, p 21; Ministry of Fuel and Power, *Efficient use of fuel* (1944), pp 92, 248–64; I G C Dryden ed, *Efficient use of energy* (1975), pp 59–63; *Materials Reclamation Weekly* [*MRW*], 14 October 1978 and 19 September 1981.
5. Manchester Steam Users' Association, *A sketch of the foundation and first fifty years' activity* . . . (1905), pp 5, 7, 23, 27, 36, 40, 72–3; *Digest of inventions for the consumption of smoke* (PP 1854 LXI), pp 22–9; Samuel Griffiths, *Guide to the iron trade of Great Britain* (1873), pp xxxviii–xxxix; Coal Commission, Committee B, *Evidence* (PP 1871 XVIII ii), qu 1166 ff; r.c. on coal supplies (PP 1904 XXIII ii), qq 12758, 14710; Babcock & Wilcox Ltd, *Steam: its generation and use* (6th British edn, 1907), p 81 and passim; *Efficient use of fuel*, p 294; Central Electricity Authority, *Report and accounts for 1956–7* (PP 1956–7 XI), p 17; Central Electricity Generating Board, *Report and accounts for 1965–6* (PP 1966–7 XXVIII), pp 13, 24, 87; ibid. for 1986–7, passim; NSAS, *Proceedings of the 18th annual conference 1951*, p 81.
6. Coal Commission, *Evidence*, qq 623, 764 ff, 1170–2, 1202; Babcock & Wilcox Ltd, *Steam*, p 133; *Efficient use of fuel*, pp 538, 541, 547; General Board of Health, *Report into the warming and ventilation of buildings* (PP 1857 XLI), pp 29–75, 97, 121–4; Department of Energy supplement to *Radio Times*, 26 June 1975.
7. A and N Clow, *Chemical revolution* (1952), pp 526 and 560; T S Willan, *Navigation of the River Weaver in the eighteenth century* (Chetham Society, 1951), esp. App. V; r.c. on noxious vapours (PP 1878 XLIV), qq 7693–8, 7758, 7798; *43rd annual report on alkali works* (PP 1907 IX), pp 61–2; *99th report* (1963), p 19; A E Musson, *Enterprise in soap and chemicals*, (Manchester, 1965), p 206; r.c. on sewage disposal, *9th report*

(PP 1914–6 XXXV), p 142; F A Talbot, *Millions from waste* (1919), p 69; H J Spooner, *Wealth from waste* (1918), pp 218–19, 223–4.

8. Coal Commission, *Report* (PP 1871 XVIII i), pp 98–101; r.c. on coal supplies (PP 1904 XXIII ii), qq 14854 ff, 15595–8, 15706; W Fairbairn, *Iron* (Edinburgh, 1861), pp 83–5; Gas Council, *1st report and accounts* (PP 1950–1 XIII), pp 17–18, 43, 78–9; M Elsas, *Iron in the making* (Cardiff, 1960), pp 182, 202; Coal Commission, *Evidence*, qu 94; s.c. on coal (PP 1873 X), App. I, p 135; Iron and Steel Board, *Iron and steel annual statistics 1956*; British Steel Corporation, *Report and accounts 1985*, p 6.
9. *Dictionary of National Biography* sub James Young 1811–83; r.c. on rivers pollution (PP 1872 XXXIV), pp 107, 110–11; r.c. on coal supplies (PP 1904 XXIII ii), qq 13346, 15067–70, 15123, 15141; Census of Production 1907, *Final report* (PP 1912–13 CIX), pp 49–50; H S Jevons, *The British coal trade* (1915), pp 694–717; *AAS*, v.d.

Overleaf:
It is rarely possible to mine or quarry minerals without a great deal of spoil. Colliery slag-heaps are well-known; slate quarries also leave a mark on the landscape.

CHAPTER EIGHT

Metals and Minerals

For its size the United Kingdom is remarkably well endowed with minerals. The variety of Britain's geology is such that a full survey of economic possibilities is out of the question here. All that can be attempted is some account of economically important metals and other minerals, like salt, of which Britain has large deposits. For most of its commercial history Britain produced and exported raw materials rather than manufactured goods. The Phoenicians came for tin, the Romans for silver, lead and iron, and Florentine merchants in the fourteenth century for wool. By the end of the eighteenth century Britain seems to have been the world's major producer of base metals – lead, copper and tin – as well as a substantial and rapidly rising maker of iron. By recent standards Britain's output of metals at the start of the industrial revolution may seem pitifully small: the output of copper and tin amounted to only a few thousand tons a year, and of lead perhaps 50 or 60 thousand tons. These tonnages, however, may have amounted to half or even three-quarters of world output. Nowadays, world output of copper and lead, and of aluminium and zinc, is measured in millions of tons. Only tin output, which has multiplied a mere fifty times, remains at the comparatively low level of about 200,000 tons a year. Iron-making was more widespread and undertaken on a larger scale. Even so, on the basis of fragmentary evidence and informed guesswork, British output of pig iron was approaching only 200,000 tons at the end of the eighteenth century, and the rest of the world was making as much again, or rather more. Some pig iron was used for castings, but most was probably turned into bar or wrought iron; the production of steel was slow, laborious and very small. World output of iron and steel has since advanced at least as rapidly as the output of base metals,* having multiplied perhaps a thousandfold in less than 200 years. Half a million tons of pig iron in 1800 contrasts with half a billion tons of steel produced today.

Britain now makes only a tiny contribution from her mines to these enormous tonnages, but kept her place as the major producer of metals until the second half of the nineteenth century. Output of copper peaked early, in 1856, at 24,000 tons, when Britain was still producing half of the world's supply. Within thirty years, the output of copper from British ores had fallen by nine-tenths while output soared elsewhere, and since then Britain's copper mines have been of negligible importance in the world's economy. The output of lead and tin reached a peak – of 73,000 tons and 11,000 tons respectively – somewhat later, in the early 1870s. The lead industry then began a slow decline, and by 1900 output had fallen by two-thirds. Tin mining fared rather better: output held up for twenty years but collapsed in the 1890s, so that by 1900 tin production from native ores had fallen to about 4,000 tons a year. It halved again by 1938 and Geevor, the last tin mine in Cornwall, has recently closed. Neither British lead nor British tin held such a predominant place in world production in their heyday as copper had done; in 1850 Britain may have produced about a quarter of the world's lead and tin. World production of these metals did not advance at the same headlong pace as with copper, so that British production still amounted to about a fifth or sixth of world output in the 1870s. By 1900, however, Britain was making only a trifling contribution, perhaps five per cent, to world output. The iron industry kept a proud place in world rankings for rather longer. In 1850 and in 1870 Britain made about half of the world's pig iron, and foreign ores were scarcely used in British works at either date. In 1889 Britain still produced a third of the world's pig iron and steel but dependence on foreign ores had begun and would steadily increase – from less than a quarter of the ore smelted in 1889 to a third in 1937. By 1987 home production of iron ore had almost ended – a paltry 300,000 tons against an import of 17 million tons. Britain's share of the world's output of iron and steel fell rapidly in the last decade of the nineteenth century to about ten per cent in 1900. Although steel production in Britain remained on a rising trend and peaked in 1973 at more than five times the level of 1900,

* Except aluminium, which was unknown in 1800. Output of aluminium now exceeds output of copper.

Britain's share of world output continued (and continues) to fall, and now stands at about two per cent.

In metal mining, the tonnage of rock that has to be moved greatly exceeds the amount of metal that can be obtained by smelting. The tin and lead ores – cassiterite and galena – may be two-thirds metal, but large quantities of worthless rock stand between the miner and the metal-bearing ore. Ores of copper and iron are much leaner than those of lead and tin, and in either case it may be and usually is necessary to remove much worthless rock in order to get at valuable ore. Even so the weight of rock that has to be moved is small compared to the enormous quantities of non-metallic minerals consumed in modern Britain. Since the Second World War the demand for chalk, clay, limestone, sand and gravel, and other minerals has more than doubled and now greatly exceeds the tonnage of coal extracted from the earth.

Table 8.1 Mineral production in Britain, 1950–87

	(*million tons*)				
	1950	*1960*	*1970*	*1980*	*1987*
TOTAL	127	192	316	287	315
Sand and gravel	39	74	108	96	111
Limestone	25	38	90	72	94
Clay, shale etc.	24	30	32	20	17
Chalk	13	15	16	14	13

It is no wonder that gravel pits full of water and brickfields stripped of clay are fairly common sights in the British landscape. There is no danger that quarrymen will level Britain's hills in the next 1,000 or the next 10,000 years, but if uncontrolled they will destroy much beautiful scenery. The waste arising each year in the British economy* roughly equals the tonnage of new materials mined, quarried or scooped from the earth. Unfortunately the waste does not necessarily occur in areas close to mines or quarries and new sites therefore have to be found for tipping. Even where quarries brickfields and gravel pits can be filled with industrial, building, or domestic waste years pass before the scars are healed.[1]

* See Chapter five.

RESERVES OF METAL ORES IN BRITAIN

The rising output curve of bulk minerals (see Table 8.1) shows that supplies are plentiful and the prospect of exhaustion fanciful, or at least distant. Metal mining in Britain on the other hand has almost come to a stop. Can we conclude that Britain's metal mines are therefore exhausted? Coal and the ores of iron are sedimentary rocks laid down in fairly regular seams. Often tilted or faulted by earth-movements they are nevertheless fairly easy to find by boring, and to mine in a systematic way. The ores of base metals, in Britain, are elusive. They exist in veins or lodes that have been extruded under pressure through other rock. When found they have often proved to be rich in metal, but the course of a vein is unpredictable and the metal-miner passes quickly from feast to famine and back again. In the eighteenth century, Cornish copper mines yielded ore containing ten or even twelve per cent metal, and ores with only eight per cent were described by John Vivian in 1799 as 'a lot of the poorest ores that I ever remember in Cornwall.' Thomas Williams, the 'Copper King' who mined Parys Mountain in Anglesey, had somewhat leaner ores, of not more than ten per cent copper. He was pessimistic about the future of the industry: no mine was inexhaustible, the existing mines were getting deeper and more costly to work; and when the Cornish and Anglesey mines were exhausted, he could suggest nowhere else to look for copper. It was certainly true that Cornish mines were getting deeper: Dolcoath copper mine was already 600 feet deep in 1790 and by 1816 had reached 1,368 feet. Tresavean mine was approaching 2,000 feet in 1830. New mines, however, often operated at shallower depths. The best-known of these was Devon Great Consols near Tavistock. For twenty years it was an extraordinarily rich source of copper, but by the 1870s its best days were over and it would have closed much earlier than it did but for the arsenic that kept it going once the copper was no longer worth working.

In the 1860s the average yield of copper from British ores had fallen to only 6.4 per cent, and smelters, mostly located in the Swansea Valley, were relying more and more on imported ores. The import trade was very small in the 1820s (about 2,000 tons a year) and remained modest in the 1840s at about 50,000 tons. Foreign ore was richer than native deposits, some of it yielding 20 per cent copper. While such rich ores were available, copper smelting could hold its own in Britain, where coal was cheap. With the inefficient and wasteful furnaces of the mid-nineteenth century it took 18 tons of coal to win a ton of copper. When the smelter needed ten or at most twelve tons of ore for every ton of copper, it clearly paid to take the ore to the coal rather than the coal to the ore. Vivian's in the Swansea Valley reduced the amount of coal needed by a quarter in the years 1856–69. The introduction of the Gerstenhoffer furnace economised on coal still further and allowed the recovery of sulphur from the copper pyrites. Such technical improvements, coupled with cheap Welsh coal and heavy imports of ore and regulus (partly refined copper) allowed the Swansea copper smelters to remain active until the First World War. In the longer run, however, it became cheaper to refine copper ore near where it was mined. Deposits of copper in the United States, Chile and central Africa took a form different from the rich but sporadic veins of ore found in Britain. Wide tracts of land were mineralised, and it became cheaper to smelt large quantities of low-grade ore than to search laboriously and expensively for rich veins. By the end of the nineteenth century it was feasible and economic to smelt ore containing only three per cent copper. In the 1950s an average grade of ore in the United States contained less than one per cent copper. Open-pit mining using the latest earth-moving equipment became the order of the day, and a huge hole in the ground has replaced the warren of shafts and tunnels that typified Cornish copper mining.

British production of the other base metals – lead, tin and zinc – has followed a similar course: reliance on native ores, followed first by imports of foreign ores and concentrates, and ultimately by the import of metal for working up into manufactures. Imports of lead concentrates had almost ended by the 1920s, but Britain still had a moderately prosperous lead-mining industry, so that up to the Second World War native ores provided about a tenth of the country's demand for lead. Zinc and tin smelters relied almost entirely on imports of concentrates, which remained substantial until after 1945 and in the case of zinc well into the 1950s. Production of native iron ore held up well until the 1960s when for the first time the tonnage of imported ores exceeded home supplies. By then mining of the rich haematite ores of Furness had practically ceased, and nearly all the British ores came from Jurassic rock yielding 26 per cent iron. Since foreign ores were almost twice as rich, it is not surprising that the steel industry sought deep-water coastal sites and took less and less British ore. By 1976 it accounted for little more than a fifth of supplies by tonnage (less by iron content), and iron mining in Britain has since virtually ended.[2]

SUBSTITUTION

The failure, or strictly the non-competitiveness, of British mines has not affected the fabrication of ferrous and non-ferrous metals in Britain. Imports of ore, of concentrates or of slab metal have more than filled the gap and the consumption of metals in British industry is now vastly greater than when Britain was the leading producer of iron, copper, lead and tin. The substitution of imported for home produced supplies has been the main way in which increased demand for metals has been met. But it is not the only form of substitution that has occurred.* Metals may replace more traditional materials; they may replace each other, one proving cheaper or more suitable than another. Or pressure on supplies of metal may be relieved by substituting other materials like glass, plastic, or ceramics. All these processes can be illustrated from the economic history of the past two centuries. Copper provides several examples. The practice of sheathing ships in copper began in Britain in 1761 with the naval frigate *Alarm.* The copper when fixed with hardened copper bolts (iron bolts caused corrosion) greatly lengthened the life of a ship. One Liverpool ship, perhaps an extreme case, reportedly had timbers as good as new in 1799, after fourteen years' service in the Africa and West India trades. Without the protection of copper sheathing the vessel would have rotted away in less than half the time, or have had to undergo extensive repairs. In effect, copper sheathing served as a substitute for timber. A cheaper method of preserving timber was to impregnate it with copper sulphate. Railway sleepers and telegraph poles were being so treated from the mid-1840s. The building of iron ships from the middle of the nineteenth century displaced therefore not only timber but copper sheathing as well. Copper, a ductile and reasonably tough metal, was used for the tubes of water-tube boilers in some warships built for the Admiralty in the

---

* Substitution is an older and perhaps more important agent of economic change than the division of labour. The substitution of bronze for stone is a well-known instance from prehistoric times. The industrial revolution is usually depicted as a time of increasing division of labour. But it can just as plausibly be described as a process of substitution: of iron for wood and stone; of cotton for wool, linen and silk; of steam-power for water- and wind-mills; of earthenware pottery for wooden and pewter platters; of towns for villages; of wage-workers for peasants and self-employed craftsmen. The list is not exhaustive, and could be matched by an equally long one for modern times.

1890s, in preference to the more usual wrought iron. In several instances the copper tubes split, and by 1900 the Admiralty was specifying neither iron nor copper but steel for the tubes of marine engines. Early in this century the use of steel rather than wrought iron became general in boilers constructed by the leading firm of Babcock & Wilcox Ltd. In more recent times aluminium has tended to take the place of copper in the cables used for high-voltage, long-distance, transmission of electricity. A good conductor, strong when alloyed, and cheaper than copper for the past forty years, aluminium had considerable advantages over the traditional metal. Meanwhile the Treasury lessened the cost of silver coins first by reducing the silver content to a half, the rest being copper, nickel and zinc; and then by issuing a cupro-nickel coin that contained no silver at all.

Before aluminium was first isolated electrolytically in 1886, it was such a rare and expensive metal that it was used in jewellery. A paradoxical state of affairs since aluminium, a constituent of clay, granite and other rocks, is the third commonest element in the earth's crust. Bulk production demands generous supplies of cheap electricity for the electrolytic process by which the metal is obtained from bauxite, the mineral from which it is easiest to detach aluminium in the present state of knowledge. World output now exceeds that of any other non-ferrous metal, and it is not surprising that aluminium has often proved a substitute for other materials. Wood and canvas were the principal materials used for the fuselage and wings of the earliest aeroplanes. By 1930 the first all-metal aircraft were in service, and a large new market opened up for aluminium. At a more homely level, the aluminium saucepan challenged cast-iron and enamelled cookware, not without some slight risk to health if aluminium salts entered too freely into food. Aluminium saucepans in their turn are now giving way, where the housewife can afford it, to stainless steel. Lightness and resistance to corrosion make aluminium a suitable substitute for iron and steel in motor-cars. Enterprising manufacturers have used aluminium instead of cast-iron cylinder blocks, and in some expensive cars aluminium body panels have replaced the traditional mild steel. Tin cans are misleadingly so called since the body of the can is mild steel thinly coated with tin on the inside. Such cans are losing market share to aluminium cans. Similarly, and to the dismay of connoisseurs of beer, the wooden beer barrel has been replaced by the aluminium keg, which is more hygienic, more durable and better fitted to withstand the pressure generated by gassy modern beers.

Metal containers, whether made of aluminium or mild steel, have now to face competition from non-metallic materials, especially glass and plastic. The glass bottle has a long history, but plastic containers are a recent challenge to metals as well as to glass. Plastic paints reduce the demand for zinc- and lead-based paints. From the point of public health any reduction in the use of lead is to be welcomed. From a conservationist point of view the change is less welcome. Oil is the feedstock from which plastics are made, and oil reserves seem likely to be exhausted long before lead-bearing ores. In this case economy in the use of metal is being dearly bought even though at present plastic emulsion paint is cheaper than lead-based paint. Just as coal is likely at some future date to become once more the dominant fossil fuel, so too in the long run lead paint may return to common use.[3]

Substitution occurs because one material is fitter for the purpose than another. Fitness may be tested by reference to one or more qualities: strength, lightness, resistance to corrosion, conductivity, ease of manufacture, price. A low price may compensate for weaker performance on other counts such as lightness, or resistance to corrosion. Conversely, a high price will discourage the use of an otherwise admirable metal. Price is also a tolerable guide when mineral reserves are under consideration. In theory all the world's untouched mineral resources are exhaustible in the long run, and if exhaustion is approaching prices will undoubtedly rise, relatively to the price of renewable commodities like wheat, or leather, or labour. The invisible hand marks up the price of scarce commodities and thus encourages the use of substitutes, which defer the day when supplies are exhausted.

METAL PRICES

There is, it would seem, little evidence in the history of metal prices that exhaustion is in sight. The resort to low-grade ores is most obvious in the copper industry, yet remarkably the price of copper has not risen unduly fast. A good series of prices for copper, tin and lead begins in 1782 with Tooke's *History of prices*. At that period the prices of tin and copper were similar, and lead could usually be bought for about a quarter or a fifth of the price of the other two. For example the average price of copper in 1782 was £85 per ton, of tin £83, and of lead £19. In 1875 tin and lead had become a little dearer relative to copper: copper was £88, tin £92 and lead £23. Zinc in 1875 cost about as much as lead. Moving on 85 years to 1960 a new snapshot gives a much changed picture. The long period of comparative price stability that began in the middle of the seventeenth century ended for Britain with the Second World War. The Revolutionary and Napoleonic Wars and the First World War had all caused inflation but had been followed by periods of falling prices so that long-run stability was maintained. Jane Austen's Dorothy Crawford with her £20,000 in 1814 would still have enjoyed a comfortable life with such a capital stock in 1938.

By 1960 creeping inflation was endemic in the British economy, and no sector escaped its influence. Copper, lead and zinc were three to four times as dear as in 1875: copper £246, lead £72, and zinc £89. Tin rose much faster and at £794 a ton was nearly nine times as dear as in 1875. Since 1960 the pace of inflation has quickened and the purchasing power of the pound by 1990/91 was about a tenth of what it was in 1960. Of our four base metals copper and zinc at £1,300 and £650 a ton have risen most but have nevertheless advanced more slowly than prices in general. Lead at £335 and tin at £3000 a ton have become relatively cheap, their price having advanced at much less than half the average rate. Over the long period from 1782 to 1990 it is perhaps fair to say that prices in general have risen to about 25–30 times their 1782 level. The price of tin, which has multiplied 33 times over the period, is the only metal among the four that has advanced faster than the general price level. Copper, lead and zinc have all lagged behind, their prices having multiplied about 15 times. The general price level, a somewhat vulnerable concept over so long a period as two hundred years, embodies a large labour cost made all the larger by the rise in real wages that has occurred. If instead metal prices are compared with the prices of renewable raw materials like wheat or wool, the pressure of demand on mineral resources becomes more evident. For, over the period since 1782, the price of wheat has multiplied only nine times, and the price of wool has merely trebled. Wheat and wool embody labour costs just like tin and copper. The slow rise in wheat and wool prices may well comfort the disciples of Malthus, but those who are concerned about non-renewable resources have less cause to rejoice.

More than thirty years ago American economists were examining relative prices for signs of approaching exhaustion of non-renewable resources. Their findings, published in an expansive and hopeful phase of world economic history, were cheerful. They reported that in mining as in agriculture firms were working under a system of increasing returns: in both sectors a given input of labour and capital yielded a much larger output in 1957 than in 1919, or than in the last thirty years of the nineteenth century. The gains were particularly striking in mining, where total productivity rose four-and-a-half times, against a doubling in agriculture. Forestry was the only extractive industry to experience diminishing returns over the period. Admittedly, somewhat less favourable findings would have been reported if the authors had tested only American data, for part of the fall in the real cost of minerals came from reliance on imports of metals from low-cost countries. Prices as well as costs reflected a favourable state of affairs. Mineral prices fell in comparison with general prices, against static prices for agricultural products and some rise in the price of timber. A more specialised study of the copper industry came to similar conclusions.[4]

But these findings are not decisive. It could be argued that the evidence of prices is misleading. Some prices have been held down by the competition of substitutes. The partial displacement of copper by aluminium and of lead-based by emulsion paint are examples already mentioned: inevitably any such substitution will tend to depress the price of the redundant metal. Similar processes are at work in other sectors of the economy. It is well known that people eat less bread when their incomes increase. If bread is no longer the staff of life it is small wonder that its price has risen relatively slowly. Wool like other natural fibres has to face the competition of synthetics such as nylon and terylene derived like emulsion paint from a petroleum feedstock. Cheap oil and cheap aluminium are among the more remarkable features of twentieth-century economic history, and their (perhaps temporary) impact on the price of a wide range of commodities makes any simple deductions from price data somewhat hazardous.

In any event the long-term danger is not to be denied. That rich deposits of minerals lie undiscovered in Brazil, China, Antarctica and elsewhere is highly probable. It is also more than possible that non-metallic substitutes for base metals may be invented, but whatever their chemical composition their manufacture will demand much energy, which itself is an exhaustible resource. That Britain's reserves of metallic ore are at present economically worthless is plain. It does not follow that future generations will not need them. However high the costs, a time may come when it will once more be profitable to mine Cornish tin and Cumbrian lead.

Cynics may think that Britain is prudently husbanding her tin, lead and coal until the rest of the world has been ransacked. But past history and the present depletion of North Sea oil suggest that neither prudence nor perfidy governs these matters. Britons will sell their wasting assets to the world for a profit, just as cheerfully as Bolivian tin magnates or Arab sheikhs. Perhaps wisely, most men give little thought to the distant future, and take what profit comes their way. Who can tell what the future holds? Only economists and conservationists claim to know; unfortunately they disagree.

REFERENCES

1. C Schmitz, *World non-ferrous metal production and prices 1700–1976* (1979), p 6; B R Mitchell and P Deane, *Abstract of British historical statistics* (1962); R Burt, 'Lead production in England and Wales 1700–1770' (*EcHR* XXII, 1969), pp 257–65; M G Mulhall, *Dictionary of statistics* (1892); *AAS*, v.d.
2. Committee on copper mines and copper trade 1799 (*Reports of committees* (PP X, 1803), pp 671–2, 679–80, 703; J R Harris, *The copper king* (Liverpool, 1964), p 42; G R Porter, *Progress of the nation* ed F W Hirst (1912), p 229; Coal Commission, Committee E, *Report* (PP 1871 XVIII iii), p 173 ff; Committee B (PP 1871 XVIII i), pp 102–3; *Account of all copper imported into Great Britain in . . . 1822* (PP 1822 XI); *Accounts of copper and tin imported . . . in the year ending 5 January 1843* (PP 1843 LII); D T Williams, *Economic development of Swansea . . . to 1921* (Cardiff, 1940), pp 79–83; O R Herfindahl, *Copper costs and prices 1870–1957* (Baltimore, 1959), pp 4, 210–11, 224; *AAS*, v.d.; E J Newell, ' "Copperopolis": the rise and fall of the copper industry in the Swansea district, 1826–1921' (*Business History*, July 1990).
3. Harris, *Copper King*, pp 45–50; G P Marsh, *Man and nature* (repr 1965), p 252 n; *Return of the number of explosions or serious leakages in water-tube boilers . . .* (PP 1901 XLII); Babcock & Wilcox Ltd, *Steam: its generation and use* (6th British edn, 1907), pp 34–44.
4. Schmitz, *Non-ferrous metal production and prices*, pp 26–8; T Tooke, *History of prices and of the state of the circulation* (1838, 1840, 1848), II, 397–420, III, 296–8, IV, 426–34; Mulhall, *Dictionary of statistics*; H J Barnett and C Morse, *Scarcity and growth: the economics of natural resource availability* (Baltimore, 1963), pp 8–9, 189, 196, 204–11; Herfindahl, *Copper costs and prices*, pp 4, 15, 203, 210–11.

Overleaf:
Waste Trade World, still going strong under another title, began life in 1912. It was the first trade paper in Britain devoted to the business of scrap and waste products in all their many forms.

WASTE TRADE WORLD.

SCRAP IRON & METALS: RESIDUES: DROSSES: COTTON & WOOLLEN RAGS: PAPERSTOCK: HIDES & SKINS:

PUBLISHED WEEKLY

WASTE RUBBER: HAIR SCRAP LEATHER FEATHERS: BONES COTTON, WOOL & JUTE WASTE: ETC

VOL. I. No. 13. LONDON; SATURDAY, JULY 27th, 1912. PRICE 1d. Annual Subscription: Home 6/- Abroad 8/-

[Registered as a Newspaper and for transmission by Canadian Magazine Post]

SECOND-HAND LINENS.
GOOD CLEAN WHITE POLISHERS & WIPERS.
CANVASSES OF EVERY DESCRIPTION.

Sails. Large Sailcloth. Tarpaulins. Painted Canvas. Large Flax, Hemp and Jute Pieces.

Every Kind of Useful Old Stores.

Government, Railway, and Steamship Stores Contractors.

Telephone {2966 London Wall. {12396 Central. Telegrams—"Outbalance, London."

JOHN PHILLIPS & SONS, Ltd., Dingley Road, City Road, LONDON.
(BY CITY ROAD BRIDGE).

Telegrams: "McKECHNIE, BIRMINGHAM." Telephones: 381 & 382 EDGBASTON.
Codes: A.B.C. 5th EDITION AND LIEBER'S.

McKECHNIE BROTHERS,

COPPER SMELTERS, METAL REFINERS : : AND MANUFACTURERS, : :

Metal Works, Rotton Park St., BIRMINGHAM.

Buyers for Prompt Cash of
SCRAP COPPER, YELLOW BRASS, GUNMETAL, PEWTER, WHITEMETALS, &c.
: : GUNMETAL BORINGS, BRASS BORINGS, : : MIXED GUNMETAL & BABBIT METAL BORINGS, &c.

LONDON: 22, Emerson St., Southwark, S.E.
NEWCASTLE-ON-TYNE: 90, Pilgrim Street.
Copper Smelting Works, WIDNES, LANCS.
Brass Ashes Works. TRAFFORD PARK, MANCHESTER.

Buyers of SCRAP IRON STEEL & OLD METALS etc.
Thos W. Ward Ltd Albion Works SHEFFIELD.

GIBSON & CO.,
118, Weston St., Long Lane, Bermondsey, S.E. LONDON,
BUYERS OF
Rabbit and Hareskins, : Horse Hair, Lambskins, &c.

Telegrams: Pass, Bristol Tel.: Nos. 3475-6

CAPPER PASS & Son, Ltd.,
BRISTOL,
ARE BUYERS OF
Tin and Solder Ashes and Tin Slags. Lead Ores, Lead Ashes, Lead Slags, :: and Sulphate of Lead. :: Old Accumulator Plates & Lead Peroxide. Antimonial Lead, Type and Stereotype Ashes, and all Dross and Residues containing
Tin, Copper, Lead & Antimony.

CHAPTER NINE
Any Old Iron?

The desire to avoid waste and the depletion of natural resources by reusing second-hand goods must be nearly as old as homo sapiens himself. There are some extravagant exceptions, well known to modern archaeologists. Egyptian Pharaohs, Chinese Emperors, Scythian, Celtic and Saxon chieftains all delighted in taking food, plate, weapons, or terracotta armies of the finest workmanship with them to the grave. In the Celtic west the habit of throwing bronze swords into lakes and rivers depleted the stock of bronze so seriously that it hastened the use of iron. These were ostentatious examples of deliberate waste. Roman armies had a more practical reason for waste when they used a javelin that broke upon impact: the enemy could not throw it back. But at a more humdrum level of society, economy rather than waste was the rule from early in man's history. In England it is possible to document the process from medieval times. There was a market on Cornhill in London for second-hand clothes in the fifteenth century. The poet Lydgate had the misfortune to lose his hood:

> Then into Cornhill anon I yode [went]
> Where was much stolen gear among;
> I saw where hung mine own hood
> That I had lost among the throng;
> To buy mine own hood I thought it wrong,
> I knew it well as I did my creed.
> But for lack of money I could not speed.

The Reformation in England put the whole stock of monastic buildings on the market. Some – Forde Abbey in Somerset, or Lacock in Wiltshire, for example – became country houses. Most, however, have come down to us as ruins, destroyed by man rather than time. Their lead, stone and timber were all reusable, and deserted monasteries remained a valuable source of building materials for a good many years. In 1544 Henry VIII's Serjeant Plumber reused lead from St Alban's and Waltham Abbeys to repair the roof of Westminster Hall. In 1550, lead from Reading Abbey was used to make a new conduit at Windsor. Elizabeth I was still selling old, apparently monastic lead in 1595.

RECYCLING

Recycling – finding a new use for goods that have already passed through the productive cycle once – is undoubtedly an old practice; only the term is comparatively new, first recorded by the Oxford English Dictionary (*OED*) in 1926, and in popular use for no more than the past twenty or thirty years. It is useful to distinguish between goods for recycling and goods that are byproducts. Recycling implies that goods have been used at least once – scrap metals, waste paper, woollen rags, cooling-tower water. Byproducts are goods that arise incidentally in business and may be wasted if no use can be found for them. Bones, scraps of leather and horn, ammonia and tar in gasworks, are all examples of byproducts. So too is the reworking of slag from an abandoned silver mine, or the recovery of grease and potash from the process of washing wool.* Finding a use for hitherto unwanted byproducts is an important source of profit for many successful firms, and like recycling means making fuller use of the world's resources. But since it is in principle a different concept from recycling (though the distinction is sometimes a fine one) the subject will be put off until the next chapter.[1]

---

* Slag from the Mount Laurian silver mines that had once provided treasure for Athens was being reworked by a French company from 1860 onwards (G Agricola, *De re metallica*, ed Hoover (1912), p 28 n). The Yorkshire woollen and worsted industry sadly neglected the valuable byproducts of wool-washing, and much river pollution ensued.

THE WASTE TRADES

Many of those who collected scrap or waste were poor and supplemented or earned a meagre living from such activities. But it would be wrong to think that the waste trades were only for the poor or flourished only in poor societies. Like other businesses they did give work on a regular or casual basis to large numbers who were poorly rewarded; the rewards for a much smaller number were considerable and the trade attracted men of substance eager for profit in exactly the same way as other trades. The prestige of a scrap-metal dealer or a shoddy manufacturer did not compare with that of a merchant or banker, but that was partly because it was unusual to make large fortunes in the waste trades. The term 'shoddy' had unfortunate overtones, and the scrap-metal business was associated in the public mind with pilfering and petty larceny. But ironmasters and paper-makers, who relied on recycled materials for part at least of their supplies, were far from social outcasts. Samuel Cunliffe-Lister made a huge fortune out of waste silk and no stigma attached to his business; far from it – he ended his days as Baron Masham. It has to be admitted that silk waste was a byproduct rather than a recycled material, and that Baron Masham strictly belongs in the next chapter. But the general public did not distinguish carefully between the various branches of the waste trades. It was in wartime and during periods of concern about the environment that reuse of scrap or waste materials – 'salvage' or 'recycling' – became a duty for the patriotic or public-spirited citizen. During a shortage of rags for paper-making in 1860 there had been a similar appeal. A worried paper-maker from Bury, Thomas Wrigley, argued that the collection of rags should be put on the footing of duty rather than profit; he described the arrangements for collecting rags as a mere barter trade between the housewife and the rag-and-bone man – 'toys and toffee, crockery and things of that sort, for old rags'. But there was no reason to frown at barter, which simply showed the search for profit operating in a somewhat unusual form. The paper-maker was forgetting Adam Smith's sardonic dictum:

> I have never known much good done by those who affected to trade for the public good. It is an affectation, indeed, not very common among merchants, and very few words need be employed in dissuading them from it. (*Wealth of Nations*, II, p 421.)

Wartime salvage and modern bottle banks are not exempt from the discipline of the market. Their success depends essentially upon popular support and goodwill, it is true, but without the cooperation of waste-trade firms the connection between public supply and private recycling cannot be made. Nor do local authorities show much tolerance for losses when they collect waste paper, bottles and other materials fit for recycling. And laudable though these campaigns are, they remain only a small part of the business of waste collection, which is in Britain largely a matter of profit-seeking private enterprise. In Germany, a country which is tidy, disciplined, well educated in the sciences and not well endowed with raw materials, concern to avoid waste has encouraged a more persistent approach to recycling, somewhat impatient of the British anxiety over profit and loss.

Second-hand goods and material – old lead, copper and brass utensils, old iron, old clothes, cotton and linen rags, waste paper and much else – arose wherever men lived and worked. An engineering works or a busy office generated a lot of scrap metal or waste paper; the large household of a rich man had plenty of old clothes, bed linen and curtains, bones, fat and swill to dispose of. A waste-trade dealer in a substantial way of business would be interested to quote for such supplies; even so, the need for local knowledge would restrict the size of firms. When it came to recovering waste from the poor a more intricate network of dealers was necessary, men and women in a small way of business who kept a shop to which the poor brought their rags, dripping and bottles. Or itinerant rag-and-bone men undertook to barter in the way described by Thomas Wrigley.

There is only limited information about these trades before the nineteenth century. It is known that plague in London in 1636 halted the collection of rags for paper-making. Towards the end of the seventeenth century the royal navy was selling the clothes of dead sailors to the trade as a modest contribution to the expense of the service. The name by which these petty traders were known changed several times, largely because of the English love of euphemism. 'Fripperers and upholders that sold old apparel and household stuff' (Kingsford, *Survey of London by John Stow,* p 199) gave way in the seventeenth century to rag-picker and old clothesman or quite commonly 'Jew old clothesman'. In the nineteenth century a popular name for the waste trader was rag-and-bone man, but many preferred the genteeler 'marine-store dealer'. At the 1831 Census only a couple of dozen traders so described themselves, but by 1851 marine-store dealers numbered two thousand,* with a corresponding reduction in the number of rag-cutters, rag-dealers and rag-gatherers. The newer term was going out of fashion within a hundred years, and has now quite disappeared. Instead the useful work that the marine-store dealer once carried out is now done by rag merchants, scrap-metal merchants and firms offering 'waste disposal services'. The trade paper, *Waste Trade World*, was renamed *Materials Reclamation Weekly* in 1969, the editor having realised that the paper was not just the mouthpiece of a trade, but the organ of a public-spirited environmental service.**

There is still a network of petty traders organising the collection of scrap. It is now rare to see a horse and cart, or even a van, driven by a mournful man with a bell, calling out 'Rags' 'Bones' or both. Such men can still be seen in London, but they are more likely to pick up articles left out on the pavement or found in builders' skips than to go from house to house collecting rags and lumber. The 'marine stores' that used to buy old bottles, rags, dripping or bones from women and children have disappeared. But scrap yards and tumble-down old sheds can be found in or near most towns, to which plumbers, builders and do-it-yourself householders can take lead, copper and other metals for sale at competitive prices.*** Men with a larger capital linked the rag-and-bone man to the manufacturer recycling scrap and waste material. In 1916 a rag-and-bone merchant wished to secure exemption from military service for one of his employees. In the course of his evidence before the London Appeal Tribunal he stated that his stock was worth £2,000:

---

* An absurd term inland, it was an accurate term in the ports, and was enforced there after 1854 under the Merchant Shipping Act.

** The trade can still laugh at itself, however. The following notice adorned the side of a hand-cart in the King's Road, Chelsea in March 1989:

Ted's Takeaway – Salvage Consultant
Rags [Harrods] Bottles [Full] Gold [24ct]
Silver – Junk – Anything

*** In June 1986 a waste dealer off Lavender Hill, Battersea, offered to buy mixed rags at £2 per cwt, lead at £9, copper at £40 and brass at £23. At the time the price of lead on the London Metal Exchange was £280 per ton, and of copper £980.

> General sensation! – to which the merchant added that he knew a man in the same line whose stock was worth £10,000. . . . He might [declared *Waste Trade World*] as easily have instanced firms whose stock may often exceed £100,000. . . . It is extraordinary how long this conception of a waste merchant as a grimy little man, with a handbarrow and a few odds and ends, takes to die.[2]

The import of materials for recycling also demanded traders of substance. The linen and cotton rags that were the raw materials of paper-making were being imported in considerable quantities in the eighteenth century, when more than a third of the paper made in Britain came from foreign rags. In 1800 imports of rags amounted to 4,000 tons, a level rarely attained again until 1814. By the 1820s imports were averaging more than 7,000 tons a year, but the growth of the trade slowed in the next twenty years: in the early 1850s imports averaged no more than 9,000 tons and contributed only a fifth of the supply. The internal trade in rags was growing much faster, thanks to the rapid increase of population and the enterprise of rag dealers. By the 1850s the makers and users of paper were becoming alarmed about the supply of the raw material: several European countries imposed heavy export duties on rags with a view to protecting their own paper-making industries; Britain in contrast was establishing free trade (including free trade in rags) between 1853 and 1860, and paper-makers watched anxiously as an export trade developed in British and imported rags. Exports had never risen above 500 tons until 1850. By 1853 they were as much as a quarter of imports. *The Times* newspaper in 1856 offered a prize of £1,000 for a successful substitute for rags, but it was several years before much white paper was made from any other material. The pioneer of esparto-grass paper, Thomas Routledge, landed his first consignment in 1859. By 1866 when the tonnage of esparto greatly exceeded the tonnage of rags imported, wood-pulp imports were still negligible. The statisticians did not separately enumerate wood-pulp as a matter of course until 1887 when it became apparent that pulp had overtaken both esparto and rags. Wood-pulp thereafter rapidly became the principal material used in paper-making. By 1913 imports were approaching a million tons; they reached 1.8 million tons in 1937 and in the 1960s rose still further. The trade in rags, the cause of so much alarm in the 1850s, grew modestly, to no more than 20,000 tons in the 1860s. It fluctuated around this level until the Second World War, and even reached the dizzy height of 44,000 tons in 1951. But for many years newsprint, most paper for books and much stationery had been made from wood-pulp. In a few hundred years time it will all have perished, and only books printed and letters written on paper made from linen rags will survive. Eventually high-quality rag paper will also disintegrate and the only original written records of British history will be the parchment rolls, membranes and books of the Middle Ages.

Woollen rags also had their uses. There had always been and still is a lively market in second-hand clothes, and the finest garments might adorn a number of wearers before ending their days as mere rags. Until the rise of the shoddy manufacture woollen rags were valued chiefly as a fertiliser, and demand was not strong enough to encourage imports. Soon after the establishment of the shoddy trade in the West Riding of Yorkshire early in the nineteenth century, an import trade grew up to supplement home supplies of woollen rags. There was at first an import duty on woollen rags because customs regarded them as woollen manufactures, and merchants found it convenient to import them 'for manure' only. In 1824, for example, imports of woollen rags (for manure) amounted to only 350 tons. By the 1850s when customs officials had recognised that woollen rags were raw materials rather than manufactures, the import of rags for shoddy had grown to over 2,000 tons. The trade then grew rapidly until the First World War, after which it declined as fast as it had grown.

Table 9.1 Import of woollen rags for shoddy manufacture

(*1,000 tons*)			
1820	0.1	1913	72
1824	0.3	1937	28
1856	2.3	1951	19
1866	16.0	1963	6
1899	33.0	1973	na

Imports of scrap metals like imports of ore were slow to develop in the nineteenth century. Scrap-iron imports in 1873, a boom year, amounted to only 20,000 tons. Britain, as one of the largest makers and users of iron, appears in some years like 1882 to have been a net exporter of scrap on a substantial scale. On the eve of the First World War imports amounted to little more than 100,000 tons, a paltry contribution to a steel output of 7 million tons. Imports fluctuated widely between the wars, and were at a high level – close to a million tons – in 1937. In the post-war world scrap iron and steel imports have run at much lower levels. Non-ferrous metal imports like those of iron were of little account before the First World War. Between the wars a somewhat larger trade developed. In 1937 for example about 25,000 tons of non-ferrous scrap was imported, mostly old brass and copper. Since 1945 the trade has grown and diversified. Not only are scrap copper, platinum, silver and other metals imported, but also 'ash and residues' for refinement and recovery. The scale of the business is not large, except by prewar standards.[3]

The sale of scrap materials often took place at auction. The first auction sales known in England occurred in 1649, and fittingly were for the sale of surplus naval vessels and prizes as scrap. Early auctions were sales by candle: the goods fell to the last bidder before an inch of candle burnt out. The origins of this strange and leisurely method of sale are obscure: it is thought that the practice was imitated from the Dutch, and that it might have derived ultimately from the ritual of excommunication 'by bell, book and candle'. However in excommunication twelve priests threw down lighted candles as sentence was pronounced – a rather different procedure. Samuel Pepys saw and was much taken with his first sale by candle in 1660; three years and a good many sales later, he was quite disillusioned with this way of disposing of the navy's surplus stores. Whether Pepys liked it or not, the new aid to business quickly became popular. Before the end of the century libraries and collections of paintings were being dispersed by auction. In the eighteenth century, land, houses and farm stock were often sold in that way. At the end of the century advertisements were distinguishing between sale by candle and sale by auction as we understand it today. The latter method was becoming much the commoner of the two, as being less tedious. Not that sale by auction was the brisk business it has since become: in 1756 the sale of the Rawlinson library (over 9,000 books) took fifty days, every book being sold separately and the auctioneer disposing of fewer than 200 lots a day; in 1803 an auction or public survey of land that sold for 6,340 guineas took nearly half an hour. Slow-moving though they were, auctions were a mild source of entertainment as well as a popular way of making sales. Whether the auction secured fair market prices is less certain: sellers were not above bidding for their own lots in order to force up the price, and buyers were suspected of forming rings to depress it. A substantial number of lawsuits ensued in the eighteenth century, and by 1826 the trade had the dubious honour of a legal textbook, Richard Babington's *A treatise on the law of auctions*.[4]

Once auctions had become a common part of business practice it did not take government long to impose taxes on them. In 1777 Lord North introduced licences for auctioneers and a duty of 3d in the pound on sales of land, farm stock, ships, and reversions in the public funds; on movables the duty was fixed at 6d in the pound. Lord North justified the tax on the grounds that auctions had multiplied of late, were mischievous to the fair trader, and were often attended with gross fraud and imposition. He added, for good measure, that even when fairly conducted they were a legitimate object of taxation. The duties were repealed in 1845, but the licensing of auctioneers remained. A tax on auctions was by no means the first intervention by the state in the regulation of the trade in second-hand goods. Beginning in 1698 a long line of statutes regulated the disposal of naval stores and later the whole trade in scrap metal. Under the act of 1698, anybody found in possession of marked naval stores without an official certificate was guilty of an offence and liable to a fine of £200. This harsh example of 'guilty until proven innocent' remained the law until partly reversed in a judgment of Lord Kenyon in 1794 (*R* v. *Banks*). He admitted evidence showing that Banks had bought naval stores from one who might be presumed to have held a certificate. The Merchant Shipping Act, 1854, required dealers in anchors and other naval stores to paint their name and the words 'Dealer in Marine Stores' on their place of business. They were to record their transactions in a book and were not to trade with persons under the age of 16, and not to cut up ship's cable for twine or paper stuff without permission and due notice. Later enactments continued to reflect public suspicion of dealers in waste and scrap. In 1861 the provisions of the Merchant Shipping Act were extended from marine stores to all trade in scrap metals. The Prevention of Crimes Act, 1871, forbade dealers from buying less than a hundredweight of lead at a time, or less than half a hundredweight of copper, brass, tin, pewter or german silver. This stultifying and unenforceable provision was repealed in the Scrap-Metal Dealers Act, 1964. But in other respects the 1964 act was more severe than its predecessors: it required all dealers, not only convicted ones, to register and keep records of their transactions. The ill-repute of the trade is well illustrated in a libel action of 1860. A Clapham rag merchant and marine-store dealer called Paris circulated a handbill in which he offered to buy dripping, bones, old clothes, rags, lead, pewter, horse-hair, bottles and much else. *The Daily Telegraph*, owned by Joseph Levy, stigmatised the handbill as an incitement to servants to rob their masters. Paris issued a writ for libel. In evidence he had to admit that no gentleman had sold him dripping or other waste, and his counsel failed to point out that gentlemen were hardly likely to take dripping and empty bottles to a marine store themselves. The jury found for *The Daily Telegraph.*[5]

Important industries depended on recycled materials. Paper-makers relied wholly on recycling until esparto grass and wood-pulp supplied alternatives to rags, hemp and waste paper. The earliest English paper mills produced a coarse brown paper for which old rope, waste paper and dirty linen rags were perfectly suitable raw materials. These were first beaten in water until the fibres had been reduced to a pulp, after which they were placed on fine wire meshes, drained, pressed and finally dried. By the end of the seventeenth century the art of paper-making had so far advanced that writing and printing papers were also being made, for which high quality linen rags were required. It was the fond hope of many an inventor to devise ways of removing ink from waste paper. A successful de-inking process would have allowed waste paper to be used as a supplement or an alternative to the limited supplies of clean rag. In 1800, Mathias Koops secured a patent for extracting ink from paper, and a joint-stock company was mooted to work the patent. Elias Carpenter, a paper-maker, foresaw considerable changes in the paper market:

> In case these manufactories shall be established and carried into effect, a very large proportion of the waste paper that is now bought up by grocers, tobacconists, cheesemongers and others, will be remanufactured, and those persons obliged to consume coarse new manufactured [paper], which will of necessity greatly increase the public revenue [because paper was subject to excise duty]. (*Commons Journal,* 1800, p 647.)

Neither this nor a later attempt in the 1870s made de-inking a reliable and profitable undertaking. It is only in recent years that a satisfactory process has been marketed. Some paper-makers now prefer de-inked paper as their raw material, but there is not enough de-inking capacity to meet the demand. If more waste paper goes for de-inking there will be less for manufacturers of millboard and cheap paper – unless the supply of waste paper is increased through more successful collection.

The collection of waste paper is a simple trade and therefore easily entered and left. Charities and local authorities have long tried to combine a little fund-raising with a useful commercial activity. As long ago as 1862 the Ragged School of London was organising waste paper collection. Armed with four trucks its boys over a period of nine months collected some 38 tons of paper, and quantities of bottles, rags, bones and scrap metal. More important but still subsidiary to the trade proper, was the intervention of local government. Before the First World War refuse usually went to the incinerator or to landfill. A few local authorities – Glasgow and the City of London among them – sorted tins and waste paper, but most confined recycling to sewage farms, the sale of street sweepings for manure and the recovery of waste heat from incinerators. During the First World War scarcity and high prices encouraged a more thrifty attitude and many local authorities became selective in what they dispatched to the incinerator. Interest in 'salvage' continued after the war and Sheffield installed plant designed to separate recoverable material from domestic waste. In London the Thames Board Mills Co. encouraged some boroughs to collect waste paper from shops and warehouses by supplying sacks for the purpose. But as a report to the boroughs warily observed:

> A large number of agencies are employed in collecting paper from business and private premises, and it is obvious that the cream is already being skimmed fairly close from this source of revenue. (*Report on . . . refuse*, 1934, pp 13–14.)

The recovery of waste paper and other reusable materials was much better organised in the Second World War than in the First. In 1940–2 waste paper was collected in larger quantities than ever before, and given the much reduced supplies of wood-pulp and pulp wood, provided half the paper-making material as against a quarter before the war. In the closing years of the war when newspapers were smaller and the output of paper of all kinds much reduced, it became impossible to keep waste paper collections at the high levels achieved in 1940–42, but they remained above 600,000 tons in 1945, close to the 1939 level. After the war local authorities, encouraged by high prices, continued to collect waste paper with a will. During the war and for a few years afterwards they supplied about a third of the waste paper sold to papermakers. The short recession in 1949 led to falling prices, and some local authorities withdrew from the business. Until 1948, when regulations under the Defence of the Realm Act were withdrawn by the Board of Trade, they had been obliged to organise waste-paper collection. Those authorities that persisted had their reward in the inflationary boom that accompanied the Korean War. The price of waste paper rose in a few months from £5 a ton to £10.10s [£10.50] and merchants guaranteed a minimum price of £6.10s [£6.50] until the end of 1954, when it was raised to £7. By 1960 enthusiasm for collecting paper at little or no profit was waning fast among local authorities. When prices rose some ten years later in sympathy with the price of wood-pulp local authorities were reluctant to make any special effort. They were directly or through waste-paper merchants already supplying mills with most of their mixed waste paper, and therefore with half (by weight) of total consumption. However, a third of the local authority supply had to be discarded as rubbish and it is likely that the prices paid by the mills to their private suppliers reflected the better quality of their waste paper. Before the Second World War waste paper supplied not much more than a quarter of the papermakers' raw material, measured by weight. In the 1960s the proportion rose to a third, and it is now about a half.

Table 9.2 Annual waste-paper supplies to paper mills

	(000 tons)	*% of total supplies (weight)*
1939	666	27
1940–5	742	49
1950–5	960	36
1960–5	1429	33
1970–5	1926	39
1980–5	1958	50

Sources: *Statistical Digest of the War; AAS* v.d.

It would be wrong to exaggerate the public-spirited appeal of recycling as a factor in the growth of waste-paper collection. It is true that during the war the public willingly set aside their shrunken newspapers and gave their old books to salvage drives. Anxiety about the balance of payments kept up the pressure for high rates of collection in the 1950s. But it was in the 1960s that a very large increase in collections took place. By then the output of paper had recovered from low wartime and post-war levels. There was therefore plenty of paper to collect, and no lack of merchants and local authorities to do the job. Their motives were more mundane than escape from national peril, or the more recent wish to save trees. Anxiety about the environment became acute only in the early 1970s and has not since abated. This anxiety has not left much trace in the waste-paper statistics. Collections went on growing in the early 1970s but fell in 1975 after the collapse of the boom of 1973–4; they then recovered, only to fall again in the depression of the early 1980s. The short-term fluctuations clearly mirrored the fortunes of the British economy. The long-term growth reflected not only zeal for recycling, but the ready availability of large quantities of paper, the relative prices of wood-pulp and waste paper, and the new techniques that made waste paper suitable for producing newsprint and stationery as well as millboard and brown paper.[6]

MANUFACTURE OF SHODDY

Paper-making relied wholly on recycled raw materials until the coming of esparto grass and wood-pulp. The manufacture of shoddy in the Yorkshire woollen industry was equally thrifty. The origins of the manufacture are uncertain and even the term shoddy has baffled those normally ingenious men, the etymologists.* In the sixteenth century Yorkshire woollens had a poor reputation: 'Flocks, chalk and other false ointments cast upon cloth is specially used in the North parts, where no true cloths are made'. (*Tudor Economic Documents*, III, p 214.)

In 1533 an official commission took evidence in Leeds and Pontefract about abuses in cloth manufacture, especially the use of flocks for the woof threads. Even at that early date Batley and Dewsbury furnished offenders out of all proportion to their size. They would have subscribed to the old Yorkshire saying that 'any fibre can be spun if it is long enough to have two ends'. At the beginning of the nineteenth century new technology offered new opportunities to the makers of low woollens. In 1801 three Scotsmen – Parker, Telfer and Afleck – took out a patent for a machine for 'preparing and reducing or teazling down clothes and other manufactured articles'. The machine, which had some affinities with the rag-beaters used in paper-making, was intended to provide flock for stuffing saddles and upholstery, and was first set to work in London. Within a few years the new rag-grinder was working in a Brighouse mill. In 1813 Benjamin Law of Howley Mill,

* Mungo, the term for a sub-species of shoddy, is etymologically just as obscure. It may be cognate with the word mongrel, meaning a dog of mixed and dubious origin. Derivation from Mungo, a Scandinavian saint's name once popular as a dog's name in Yorkshire, is a more fanciful suggestion.

Batley dispatched thrums and rags to Brighouse for grinding, and then spun the resulting 'wool' or shoddy into yarn. Traditionally this marks the start of shoddy manufacture, though it could have begun a year or two earlier. A Leeds manufacturer giving evidence before a House of Lords committee in 1828 thought that the manufacture had begun ten or fifteen years before. The term shoddy was evidently unfamiliar to their lordships because as a Bermondsey woolstapler explained: 'The term the manufacturers give it [cloth, carpet etc. made from rags] is "shoddy".* When a broom is used upon it, it keeps wearing out; they are goods made for sale and not for wear.' (HL Committee . . . British woollen trade, 1828, p 86.)

A firm of the highest repute, Benjamin Gott, did not use rags in the manufacture of blankets, but it did buy cheap goods containing rags, for sale to Red Indians. The shoddy trade was clearly well established by 1828, and it went on growing until the First World War, as rag imports testify. It was recognised by 1828, and perhaps from the start of the trade, that shoddy wool did not make saleable cloth unless mixed with some yarn made from new wool. The quality of shoddy wool varied – in 1828 the rags from which it was made might cost as much as £25.10s or as little as £7.10s per ton. Soft woollen goods such as stockings and flannels – goods that had been neither milled nor felted – were used to make shoddy, the best of the recycled wools. Hard-spun and felted cloth such as broadcloth became mungo. Mixture cloths containing cotton or linen were first submitted to the process of carbonisation – the destruction of the cotton or linen fibres by hydrochloric or sulphuric acid. After grinding (otherwise called 'pulling') the resulting wool was known as 'extract', an inferior commodity mostly exported to Europe without more ado.

A group of small towns south-east of Bradford – Dewsbury, Batley, Ossett and Morley – became the area of shoddy manufacture, probably because of their long association with the making of cheap woollens. Auction sales of rags were being held at Batley railway station as early as the 1830s and by the 1850s there were weekly auctions of shoddy wool, evidence of a well established and flourishing trade. The growing import of rags appears to have been matched by a similar growth of rag collections at home, and early in the twentieth century the consumption of shoddy wool amounted to 180 million lb, or as much as a third of the total

* The earliest reference given in *OED* (1989) is dated 1832.

supply of wool to the industry. In the woollen, as distinct from the worsted, industry the use of shoddy wool had become common practice, and some shoddy was used even in the manufacture of the finest west-country cloths. A shortage of new wool caused by drought in Australia contributed to the high consumption of shoddy wool in the early 1900s. The import of new wool later recovered, and although rag imports went on increasing until 1913 the proportion of shoddy to new wool fell back to about a quarter. Between the wars, when the cotton industry entered upon a long period of decline, the woollen and worsted industry was rather more successful in holding on to its markets, domestic and foreign. Exports fell off after 1929, mills closed and the labour force shrank. Nevertheless output held up and as late as 1950 consumption of new wool was as high as it had been at the beginning of the century. The lower end of the trade suffered most, and by 1950 mungo and shoddy supplied only 71 million lb, an eighth of the wool consumed. The trade held up reasonably well at this lower level of activity until 1958 when a more rapid decline set in.

The manufacturers could deal well enough with mixture fabrics containing cotton or linen, by the process of carbonisation. Fabrics containing synthetic fibres like nylon or terylene defied their best efforts: the intrusive synthetics were hard to identify – rag-sorters who used a microscope would not sort many rags in a day; and when found, they were difficult, if not impossible, to separate from the reusable wool. The use of man-made fibres in the woollen and worsted industries had scarcely begun in the 1950s, but then advanced rapidly until in the mid 1970s the consumption of wool exceeded that of synthetics by little more than a third. In the 1980s synthetics fell back somewhat, but still provided more than half as much fibre as wool. When first introduced synthetic or partly synthetic fabrics were expensive; now it is the all-wool garment that is the expensive luxury good. When synthetics are cheaper than goods made from new wool, who wants goods made from shoddy? The production of mungo and shoddy halved between 1960 and 1972 and is now trifling. A thrifty episode in the history of woollen textiles has ended.* If shoddy manufacture ever revives it will owe its return to the exhaustion of the feedstock, largely oil, from which the newer synthetic fibres are made.[7]

---

* Even the dust from rag-grinding had its uses: it was sold in the nineteenth century as fertiliser.

Table 9.3 Wool, shoddy and synthetics: consumption in the woollen and worsted industries

		(*million lb*)	
	Wool	*Shoddy/mungo*	*Synthetics*
1950	518	71	15
1960	481	68	77
1967	360	54	133
1976	277	32*	201
1987	240	–	130

Source: *AAS* v.d.

* 1972, the last year of the series

WASTE HEAT RECOVERY

Recycling rags, once the necessary machinery has been invented, is a straightforward process; recycling energy, as the laws of physics decree, is ultimately impossible. It is true that low-grade energy – for example in slightly warmed air above a power-station chimney – is of no economic value; but in higher concentrations energy is as susceptible to recycling as rags, paper, or scrap metal. For example, blast-furnace gases have been used to heat coke ovens, and waste heat from boilers to pre-heat feed-water in an economiser. The electricity industry has experimented with fish farms that take advantage of the large quantities of lukewarm water available at power stations. Another source of energy is to be found in domestic refuse. In the nineteenth century, indeed until the decline of the coal fire set in between the wars, cinders and ash accounted for over half of the refuse in dustbins. In 1893 a report to the London County Council estimated that cinders, breeze (small cinders) and ash constituted 64 per cent of London dustbin refuse. In Sheffield after the First World War the corporation arranged for the screening and separation of domestic waste: it was found that between 35 and 45 per cent was cinders and about 30 per cent was dust, no doubt mostly ashes. In Manchester and London in the 1930s the competition of gas and electricity for heating and cooking was reducing the quantities of cinders and ash to no more than half the domestic waste. By 1970 ashes and cinders formed only a quarter of domestic waste, and in London by 1980 this form of refuse had almost disappeared, the triumph of other forms of heating over the coal fire being by then complete. Elsewhere the coal fire retreated rather more slowly.

While it lasted the presence of cinders in domestic refuse gave opportunities for recycling energy. The sieving and sorting of refuse by hand was an unpleasant job, often done by women, in Victorian dustyards. A long article in the London newspaper *The Standard* described a huge dustyard near Vauxhall in 1871. There were three kinds of waste, strictly separated: stable manure, street sweepings, and house refuse. The house refuse was sieved and sorted to separate ashes, cinders, old crockery, paper, rags, bones, glass both white and dark, and iron, brass, corks, even string. Hardly anything from this miscellany was wasted, not even the broken crockery, which served as hardcore in road-making. The cinders and ash found a ready market among brickmakers, the ash as a constituent of bricks and the cinders providing part of the fuel for firing them. As the nineteenth century drew to a close private contractors gradually gave way to municipal dustmen. The sorting of refuse went out of favour and incineration often took its place in larger boroughs. They collected enough refuse to make incineration an alternative to landfill and one that could be expected to make worthwhile returns.

The first incinerator or 'dust destructor' was marketed in 1875 by its inventor Albert Fryer, and like many attempts at using waste encountered the predicament expressed in the phrase 'recycling good, pollution bad'. Slow combustion of the contents of privy middens – kitchen waste and night soil as well as cinders and paper – proved to be a thoroughly noxious process. It was not long before other inventors proposed forced draught and heated blast as an aid to combustion, and improved destructors were shown to produce large quantities of heat that could be used to pump sewage or generate municipal electricity. Temperatures of 3,000°F were not unknown and clinker containing fused glass and crockery made a particularly strong though rather dark cement. High temperatures and high chimneys did much to abate, without entirely eliminating, the nuisance caused by incineration. By 1914 most large towns disposed of part at least of their refuse in dust destructors: of 338 incinerators then at work 295 were fitted with boilers for heat recovery. Between the wars incinerators went out of fashion: the plant worked under arduous conditions and was expensive to maintain. The quality of the free fuel, never high, diminished as cinders gave way to paper and cellophane wrapping in the dustbins; and cheap electricity became available from the national grid. As plant wore out it was scrapped, and boroughs chose landfill as their preferred method of waste disposal. If old incinerators were replaced it was without heat recovery apparatus.

In the past twenty years opinion has turned away from landfill, itself a form of pollution, and interest in ways of burning domestic and trade refuse has revived. It has been estimated that the 30 million tons of trade and domestic refuse collected each year contain the equivalent energy of twelve million tons of coal. Much of that potential energy would no doubt be used in separating the non-combustible and drying the putrescible (vegetable peelings etc.), but enough energy ought to remain to make its recovery worthwhile. A small part of London's refuse (400,000 out of nearly three million tons) is burnt at an Edmonton plant that yields a net income of £1 million from the sale of electricity. East Sussex County Council has produced a fuel in pellet form from domestic and trade waste. It is easy to handle, non-polluting, and cheap, and has a fair calorific value – 9580 BTU per lb, three-quarters of the energy potential of good coal. In Manchester the waste disposal authority has encouraged the development of a process for converting domestic waste into crude oil. None of these schemes has greatly changed the pattern of waste disposal, for landfill remains the destination of nearly all Britain's waste.[8]

Landfill, though an unlovely process, has had its uses. Manchester Corporation acquired Chat Moss and Carrington Moss west of the city in 1893, and by 1914 had tipped 1.7 million tons of refuse there, with complete reclamation of the moss lands as their reward. The use of London's refuse as landfill at brickworks had become a regular practice by the 1930s and as was noted in Chapter five the brickfields of the South Midlands now accept refuse from as far away as Bristol on a systematic basis. There are more than thirty sites in Britain where landfill and reclamation are not the only purposes in mind. It is also intended to extract and burn the methane that is generated in the decay of refuse. A tip will begin to give off a steady supply of methane within 6–18 months, depending on temperature and wetness. In Britain, tips have a half-life of two to four years; that is, half the gas will be given off in that period, half the remainder in the next period and so on. The National Coal Board was a pioneer in the exploitation of this unexpected natural resource. In the 1970s it gained experience in generating electricity from methane occurring in its coal-mines, and in 1978 it devised a scheme for Aveley in Essex. Aveley was the site of a huge tip for London refuse close to paper mills at Purfleet. A network of shafts and pipes laid down almost as if for land drainage collected gases which were 55 per cent methane. A flow of 3,000 cubic metres of gas per hour supplied a third of the fuel consumed by the paper mill. Shanks & McEwan propose a similar scheme for their site at Calvert in Buckinghamshire, with nearby brickworks as obvious customers. In 1988, 36 privately owned power stations were fuelled by landfill gas; by 1992 the output of gas from such sites had nearly trebled. If all the 400 suitable sites were tapped for gas, they might save three or even five million tons of coal a year. The schemes already operating have shown what is possible without making more than a slight, though welcome, contribution to energy supplies.

GLASS AND METALS

In principle any process of recycling saves some energy since the goods recycled – paper, rags, metal, glass – embody the energy required for their manufacture in the first place. Glass and scrap metals in particular embody a great deal of energy and to recycle these goods therefore saves more than raw materials. Victorian rag-and-bone men and marine-store dealers were avid collectors of bottles, which they resold to bottle warehouses. Many brewers and mineral-water manufacturers encouraged the return of bottles by taking a deposit on every bottle, and cider-makers were still levying a deposit on flagons less than thirty years ago. During the First World War when new bottles were scarce it became more important than ever for brewers to recover their old bottles. London brewers had a bottle exchange that operated in an area stretching from Brighton to Ipswich. The Southampton brewers had an exchange of their own and an effective one, for Clayton's Brewery recovered 997 out of every 1,000 bottles. When faced with a marine-store dealer who was selling named as well as unnamed bottles to the Teign Cider Co, the Southampton bottle exchange resorted to law. *Barlow* v. *Hanslip* was decided in favour of the bottle exchange, and Hanslip, the offending dealer, was restrained from selling named bottles. Since brewers and others guarded their ownership of bottles so jealously, and dealers were willing to buy, it is not surprising that relatively few sound bottles went into dustbins. In 1919, it was estimated, less than one per cent of domestic refuse (by weight) was glass. By 1930 carelessness was creeping in: three per cent of Manchester's refuse was glass.

The Second World War enforced thrift again and in the 1940s only two per cent of refuse was glass. As plenty returned, fewer and fewer businesses encouraged the return of bottles, and more and more glass found its way into dustbins. In the 1960s glass constituted at least five per cent of domestic refuse, and by 1984 eight per cent. Even higher figures were recorded in London where in 1979 twelve per cent of the refuse was glass. The first reaction of outraged environmentalists was to dump hundreds of bottles on the doorsteps of brewers and other sinners against the light. A later and more constructive response was to organise bottle banks. A bottle bank is a less effective form of recycling than a system of deposit because it supplies cullet or broken glass rather than reusable bottles. It has the advantage, though, of finding a use for the profusion of glass jars that pickle and jam makers had never wanted back. It is probable that jars and wine bottles form the greater part of the glass dropped into bottle banks since beer is now more often sold in cans.

Glass manufacturers became eager to recycle jars and bottles in the aftermath of the high fuel prices established in 1973. Bottle banks were already known elsewhere in Europe when in 1977 the Glass Manufacturers' Federation set up its first banks. Sited in local authority and supermarket car parks they attracted a sometimes embarrassing amount of glass, and have earned modest sums for ratepayers, or charities. The London Borough of Richmond, for example, made a clear net profit of £130,000 over a period of nine years from the energetic collection not of glass alone, but also of paper, cans and rags. Bottle banks must be useful – otherwise glass manufacturers would not promote them. But they have been superimposed on an existing order of recycling. In 1976 before there were any bottle banks at all, a quarter of new glass was already made from cullet, much of it arising in glassworks themselves. Later figures that exclude cullet arising in glassworks show that the amount of glass recycled doubled between 1983 and 1988. At the latter date 275,000 tons of cullet was being collected. By 1990 there were over 4,300 bottle banks and the industry was confident that 5,000 would be in place well before the planned date in 1993. Despite this substantial achievement, Britain recycled much less of its glass than most countries in the European Economic Community.[9]

HAZARDOUS WASTE

It has become customary to distinguish between waste and hazardous waste. The distinction is somewhat artificial. Ordinary domestic waste when carelessly tipped may attract rats and flies, give off smells, and even cause explosions if methane seeps into a confined space. Hazardous waste simply needs more careful handling. It may legally be tipped under licence. It is sometimes best incinerated. It is sometimes best recycled. In all these respects it is like ordinary waste, but added precautions are needed. It can be plausibly argued, indeed, that the main difference between ordinary and hazardous waste is that hazardous waste is more profitable – where there is danger there may be profit. The best known examples of hazardous waste that arise at home and are also imported are the nuclear fuel rods reprocessed at Sellafield in Cumbria. From the beginning of nuclear power it had been realised that recyling of spent rods offered scope for profit since the rods and the byproduct plutonium were too valuable to be wasted. In other industries residues were less valuable, and ignorant or careless management often missed the chance to take another profit from whatever was worth separating out from the waste. In their defence it could often be said that one firm's residues might not provide enough material for economical reprocessing.

Since hazardous waste is an indefinite term it is not surprising that the quantity of such waste is unknown. In 1966, according to an estimate by the royal commission on environmental pollution, British industry generated about 600,000 tons of acid, caustic, or toxic waste. In the mid-1980s the chief inspector of hazardous waste estimated that about four million tons of hazardous waste had to be disposed of each year.* It is difficult to reconcile these widely divergent figures. There has probably been some genuine increase in the amount of hazardous waste produced, but it seems likely that much of the difference is accounted for by an under-estimate in 1966 and by changing definitions of what constitutes a hazard. In any case, these large quantities provide some opportunities for recycling, either by specialist firms or through facilities installed by producers of hazardous waste.

Some of the early contracts secured by Re-Chem Ltd illustrate the scope for savings open to an alert business. Re-Chem was

* Nearly two-fifths of the hazardous waste was classified as special, i.e. dangerous or difficult to dispose of.

founded in Southampton in 1967 by a chemist and an accountant. One of its early coups was to pay a company ten shillings (50p) per ton for 50 tons of effluent a week. Previously Re-Chem's supplier had had to pay a contractor £200 a week to cart it away and dump it. The effluent was rich in sulphuric acid, which Re-Chem recovered and sold at a profit to a fertiliser manufacturer. They went on to recover caustic soda, copper, and ferric chloride from other effluents. The plating industry supplied toxic cyanide solutions from which Re-Chem recovered gold and silver. Waste unsuitable for reprocessing they disposed of by high-temperature incineration. Within a few years the firm could treat 50,000 tons of hazardous waste per annum in plant at Fawley near Southampton and at Pontypool. This represented more than half the national capacity for the incineration of hazardous waste, a capacity which in total amounted to 80,000 tons in 1987. The quantities actually treated – 71,000 tons, of which 8,000 tons were imported – did not leave a large margin for future growth of the trade. A similar tonnage of hazardous waste was disposed of at sea, either by incineration or by dumping. But most – 83 per cent – went to landfill at licensed sites. These were chosen, so the public was assured, in such a way that there was no danger of pollution either of the atmosphere or of water supplies. Only time will tell whether these assurances are worth as much as the paper they are written on.[10]

HIGH-TECH SCRAP COLLECTION

The success of firms like Re-Chem shows the extent to which advanced technology has impinged upon the waste trades. The process has been long and slow. Before the First World War the London Electron Works in Limehouse, owned by the German firm of Goldschmidt, was recovering tin from cans and tin-plate clippings by an electrolytic process. In 1912 the electro-magnet was being recommended as an easy way of separating iron from other waste. The principle of the electro-magnet had been hit upon by William Sturgeon in 1823, but a practical machine was still a comparative novelty in Britain on the outbreak of the First World War. After the war the mechanical sorting of waste became common, and the humble and dirty work of sorting by hand gradually ended, or was simplified by the use of electro-magnets and of conveyor belts with wire meshes for sieving out dust and other small items. In some systems a forced draught was used to separate paper and other light material. The borough of Eccles found that separation greatly increased the efficiency of its steam-raising plant: whereas in 1914 10,000 tons of unsorted refuse evaporated two million gallons of water, in 1923 the cinders from a similar weight of refuse evaporated more than twice as much. Scrap-metal dealers handling small quantities of lead, copper and other metals brought to them by plumbers and householders needed little equipment beyond a shed and a weighing machine.

Larger firms that dismantled derelict factories or handled hundreds of scrap cars every week were effectively part of the engineering industry rather than petty traders. They had at their disposal an ever-widening array of machines and instruments. For example in 1966 the latest machinery included mobile cranes capable of lifting 5.5 tons; a 38-ton scrap-metal baler; a paper baler that handled three tons of paper an hour; 'iron shark shears' for cutting up bulky metal waste; incinerators that complied with the Clean Air Act, 1956; and portable power saws of American design. In the same year, the Japanese announced a machine, the 'carbecue', for the efficient disposal of scrap cars. After removal of its engine and tyres the car entered a pre-heating chamber that reduced upholstery and paint to ashes; in the 'carbecue' chamber the heat rose gradually from 200°C to 1,600°C, melting successively the lead, aluminium and copper, which ran off and were collected. The shell was then fed into a scrap press and crushed, before recycling into steel of high purity. Early models of the carbecue disposed of 300 cars a month, but a larger model with a throughput of 1,000 cars was planned. In 1975 Proler Cohen, a subsidiary of an engineering firm known as the 600 Group, boasted of its 'huge appetite for old bangers – so pass along the car please'. The plant received cars from within a 100-mile radius. The scientific-instrument maker also had a role to play in the work of the avant-garde scrap-metal merchant. In 1988 Arun Technology of Horsham marketed an aluminium-alloy analyser, weighing 18 kg and optimistically described as portable. It was a spectroscope that with two minutes of unskilled labour, a carbon electrode and some electricity, recorded the percentage of aluminium in a sample, and also gave readings for silicon, iron, copper, magnesium and several other metals including chromium and titanium.

Not all firms in the waste trades availed themselves of these technological and scientific wonders. In 1978 about 10,000 firms were registered under the Old-Metal Dealers Act. Many, perhaps most, were in a small way of business, probably lineal descendants of rag-and-bone men, and in the snobbish opinion of the organ of the trade they could 'hardly be classified with the reclamation industry as such'. Certainly, few of them would need an aluminium-alloy analyser. About 900 firms belonged to the British Reclamation Industries Confederation (BRIC). The average member of BRIC employed about 20 men, and had plant and transport worth no more than £12,000 per employee; about a fifth of this sum represented transport, the rest machinery and buildings. These statistics depict an industry not highly capital-intensive, but well-equipped in view of the smallness of the average firm. A short account of one forward-looking non-ferrous-scrap-metal merchant completes this brief technical account. In 1972 the Leeds branch of the firm, A A Bramall Ltd, had hardly any machinery – a forklift truck, a Lindemann baler and a set of small shears; all this equipment was very old. The firm's records were computerised in 1985. In 1986 it built a 21,000 sq ft warehouse on the site of an old clothing factory, and two years later added a 6,000 sq ft shed for sorting aluminium scrap, and a brass conveyor system. Other new plant included two copper balers, an aluminium baler and several sets of shears. The firm handled small parcels of scrap – in the year ending September 1987, 17,000 consignments were delivered by customers, weighing in total 10,000 metric tons, or rather more than half a ton per transaction. It was in such ways that a traditional industry slowly evolved to take advantage of late twentieth-century techniques.[11]

SCRAP AND TOTAL SUPPLY OF METALS

Although old metal has always had a value, recent environmental anxieties have made some difference to the amount of scrap collected. There are no reliable statistics for the supply of scrap to metal producers for earlier centuries, though it is known that old lead brass and copper made a useful contribution to supplies. In the 1930s it was estimated that about a third of the output of non-ferrous metals came back as scrap for reprocessing. Recovery rates in recent years have been a good deal higher for lead and copper, and the reprocessing of tin is now no longer exceptional, but general.

Table 9.4 Recycled metal as a proportion of supply (%)

	1974	*1988*
Lead	65	70
Copper	40	45
Tin	12	60

Source: *Social Trends 1991*

The consumption of aluminium before the Second World War was small (less than 50,000 tons p.a.) and exceeded 100,000 tons for the first time in 1940, the year of the Battle of Britain. It was in that year that Lord Beaverbrook, Minister of Aircraft Production, appealed to housewives to give their aluminium pots and pans for manufacture into aircraft. The appeal had some success, and it was no mean feat to have collected in 1940 as 'scrap' nearly as much aluminium as had been manufactured in 1938, the last full year of peace. After the war aluminium scrap normally supplied about a third of the annual consumption, rising to two-fifths in the 1970s. Since then, perhaps because of the increasing use of aluminium cans, the proportion of the metal recovered as scrap has fallen to a quarter. The recent belated spread of can banks may in time reverse the downward trend.

In steel-making, scrap provided more than half the charge of metal (the rest being pig iron) at most dates between the end of the First World War and the acute depression in the industry in the early 1980s. In recent years scrap has provided rather less than half the charge because of a sharp reduction of scrap arising in the works. This reduction is much greater than the fall in steel output and reflects improved working practices rather than a failure to use the scrap available. For many years between a third and a quarter of crude steel production had remained in the works as scrap arising, for example when billets bars and plates were trimmed before delivery. A similar wastage occurred in the production of non-ferrous metals, with the result that producers always had some scrap available in their own works ready for recycling. Modern engineering practice requires the shaping, often elaborate shaping, of pieces of metal. The offcuts and shavings (or swarf) were scrap of known composition, and copper refiners often stipulated that such process scrap should be resold to them. In the absence of any such requirement the scrap-metal merchant was a likely intermediary in the disposal of swarf. Before the Second World War contaminated scrap-copper and metal residues were exported because Britain had no facilities for electrolytic refining, a deficiency since made good. Two-thirds of scrap copper originated in the 1930s as process scrap, a much higher proportion than in the United States where other sources of scrap were more thoroughly exploited.

The scrap-metal merchant had nothing to do with scrap arising in a steel or copper works, and only a limited involvement with process scrap. His business was the collection of old metal: it was a large business – in 1977 members of the British Scrap Federation, itself a member of BRIC, handled nearly 11 million tons of ferrous and non-ferrous metal (but this figure does include some double counting). Some of this large tonnage found markets abroad, for there had long been export as well as import of scrap.[12] It is clear that scrap recovery makes a valuable and growing contribution to supplies of metal, glass and paper, and to that extent depresses demand for virgin raw material. However, as with energy supplies, economy leads in the end to a wider market. It would be a bold economist who asserted that in the long run scrap recovery would save the softwood forests or prevent the depletion of mineral reserves. The use of scrap simply gives man more time to make other arrangements for the conduct of the world's economy before reserves give out.

REFERENCES

1. R Bradley, *The passage of arms* (Cambridge, 1990); W W Skeat, *Specimens of English literature 1394–1579* (1887); H M Colvin, *History of the King's works* (1975), III, pt I, pp 37, 89, 175; *OED*, 1989 edn; D Woodward, 'Swords into ploughshares' (*EcHR* XXXVIII, 1985), pp 175–91.
2. S.c. on paper (export duty on rags)(PP 1861 XI), qq 470, 824; A Smith, *Wealth of nations*, ed Cannan, II, p 421; H A Harben, *A Dictionary of London*, (1918); C L Kingsford, *Survey of London by John Stow* (1908)p 119; *Calendar of Treasury Books* V pt i (1676–9), 21 March 1676, and VI, 26 May 1680; *Waste Trade World* [*WTW*], 9 January 1937, 13 August 1966; Old Metal Dealers Act 1964 (1964 c.69); D C Coleman, *The British paper industry 1495–1860* (Oxford, 1958), p 37; *WTW*, 2 December 1916.
3. Coleman, *Paper industry*, pp 89 ff, 109; *An account of the . . . rags imported into . . . and exported from the United Kingdom* (PP 1854 LXV); *History of The Times 1841–1884* (1939), p 213; *Annual statement of trade* v.d.; *Trade and navigation accounts* v.d.; r.c. on rivers pollution, *1st report*, p 19, *Evidence* II qu 2237 ff (PP 1866 XXXIII); for early imports of woollen rags there are returns in PP 1822 XI and 1843 LII; House of

Lords committee on the state of the British wool trade (PP 1828 VIII), p 336 App 2; *Overseas trade statistics of the UK* v.d.

4. *Calendar of State Papers Domestic 1649–50*, pp 170, 529, ibid. *1651*, p 296, *1651–2*, p 34; S Pepys, *Diary* (1970), ed R Latham and W Matthews, 6 November 1660, 28 February 1661, and 28 September 1663; J Evelyn, *Diary* (Oxford, 1955), ed E S de Beer, vol 4, pp 126, 330 n, vol 5, pp 144–5, 206; *Autobiography of William Stout of Lancaster 1665–1752*, ed J D Marshall (Manchester, 1967), pp 139, 149, 193; *The Purefoy letters*, ed G Eland (1931), I, 109–10; A Everitt, 'The English urban inn 1560–1760' in ed A Everitt, *Perspectives in English urban history* (1973), p 107; F Hermann, *Sotheby's: Portrait of an auction house* (1980), p 6; J R Walton, 'The rise of agricultural auctioneering in eighteenth- and nineteenth-century Britain' (*Journal of Historical Geography*, January 1984, p 33; *The Postmaster*, Exeter, 10 November 1721, 16 February 1722; *The Gazetteer and New Daily Advertiser*, 27 January and 27 September 1773; S Johnson, *The Idler*, no 35, 16 December 1758.
5. Parliamentary History XIX cols 246–7; 9 William III c 41 (1698); R v Banks 1 esp. 144 (*English Reports* vol 170, pp 307– 8); 17 and 18 Vic c 104, s 480 ff (1854); Paris v Levy 2 F and F 71 (*English Reports* vol 175, pp 963–66).
6. Coleman, *Paper industry*, pp 26–32, 62–3; *House of Commons Journal* 1800, vol 55, p 647; P L Simmonds, *Waste products and undeveloped substances* (1873), pp 30 ff, 273; *Materials Reclamation Weekly* [*MRW*], 8 October 1988; *The Independent*, 7 May 1990; *Municipal yearbook for 1914*, *for 1920–1*, pp 507–8, *for 1922*, p 559 ff; Metropolitan boroughs standing joint committee, cleansing sub-committee, *Interim report on disposal of refuse* (1934), pp 13–14; *Municipal yearbook for 1949*, pp 1136–44, *for 1950*, pp 146, 156, *for 1951–2*, pp 141–4, *for 1956*, p 128; *MRW*, 21 February, 19 December 1970, 18 January 1975.
7. Leake's treatise on the cloth industry 1577 in eds E E Power and R H Tawney, *Tudor Economic Documents* (1924) III, p 214; Maud Sellers in *Victoria County History: Yorkshire* (1912) II, p 411; S Jubb, *History of the shoddy trade* (1860), pp 17–19; J Bischoff, *A comprehensive history of the woollen and worsted manufactures* (1842), II, 175, 179–82; House of Lords Committee on the state of the British woollen trade (PP 1828 VIII), pp 86, 139–44, 179, 206–18, 246–8, 281–3; W S Bright McLaren, *Spinning woollen and worsted* (1889), pp 187–8; Jubb, *Shoddy trade*, pp 34, 39; H Burrows, *History of the rag trade* (1956), pp 33–4; J H Clapham, *The woollen and worsted industries* (1907), pp 26, 120–5; *Census of production 1907: final report* (PP 1912–3 CIX), pp 295, 342; *MRW, 1971 handbook and buyer's guide*, p 219 ff.
8. H J Spooner, *Wealth from waste* (1918), p 272; *Municipal yearbook for 1921–2*, p 539 ff; Metropolitan boroughs, *Report on disposal of refuse* (1934), p 6; W L Hall, 'Methane from waste', NSCA, *52nd annual conference 1985*, pt I p 3 ff; Simmonds, *Waste products*, pp 440–8 quoted at length from the article 'Dust ho!' in *The Standard*; W Robertson and C Porter, *Sanitary law and practice* (3rd edn, 1908), pp 106–15; T Koller, *Utilisation of waste products* (3rd English edn 1918), pp 7–10; W Short 'Air pollution control – incinerators' (NSCA, *51st conference 1984*), pp 1–3; East Sussex Enterprises Ltd, *Effective waste management* (1988); *The Times*, 16 September 1989; *Municipal yearbook*, v.d.

9. Redford and Russell, *Local government in Manchester* II, 428; *Municipal yearbook for 1914*; Shanks & McEwan, *Transfer and disposal of local authority waste* (*c.* 1988); *The Independent*, 17 December 1990; Hall 'Methane from waste' (NSCA *1985*), pp 9–11, 16, 19–20, 27–8; *WTW*, 26 February and 4 March 1916; W Short, 'Air pollution control' (NSCA *1984*), p 4; r.c. on environmental pollution (PP 1970–1 XVI), *1st report*, p 13; *Municipal yearbook for 1949*, p 1136; F A Talbot, *Millions from waste* (1919), pp 143–5; J Gordon, *Sharing resources* (Local authority recycling advisory committee 1988), pp 11–27; *MRW*, 9 April 1977, 26 September and 10 October 1981, 8 October 1988; *Social Trends 1979* and *1991*.
10. R.c. on environmental pollution (PP 1970–I XVI), p 14; D A Mills, 'The role of the Hazardous Waste Inspectorate' (NSCA, *51st annual conference*), pp 1–5; *MRW*, 4 July 1970, 18 January 1975, 9 July 1988; Spooner, *Wealth from waste*, pp 238–9; Hazardous Waste Inspectorate, *first report* (HMSO, 1987); *second report* [on 1985–6] (undated).
11. *WTW*, 4 May 1912; Talbot, *Millions from waste*, pp 226–7; *Municipal yearbook for 1925*, pp 706–10; *WTW*, January and February 1966, 20 August 1966; *MRW*, 4 January 1975, 25 June 1988, 14 October 1977, 5 March 1988.
12. Burt, 'Lead production in England and Wales 1700–1770' (*EcHR* XXII, 1969), p 261; Woodward, 'Swords into ploughshares'; *WTW*, 20 November 1937, 18 March 1944; *Statistical digest of the war* (HMSO, 1951); *AAS*, v.d; *Iron and steel annual statistics*, (1956); Iron and steel corporation of Great Britain, *Report to 30 September 1951* (PP 1951–2 XVI), pp 17–20; *WTW*, 30 July 1966; *MRW*, 14 October 1978.

Overleaf:
The alkali manufacture is the classic case of an industry that for many years polluted air and water because it failed to find a use for its byproducts. Later it did find ways to use them, partly out of self-interest, partly under pressure from government inspectors.

F.W.TINKER
SOAP
MANUFACTURER
TO BE SOLD

CHAPTER TEN
Byproducts

Byproducts are as old as hunting. Primitive man hunted for food, but took care not to waste the inedible parts of the animal. He clothed himself in the skin, made pins and ornaments from the bones and horn, and used stag's antlers for a pick (for example in the flint mines at Grime's Graves in Norfolk) and probably as a primitive tool for cultivating the soil. The lessons then learned have not been forgotten, and economy in the use of incidental materials has an important place in modern industry. A byproduct, as the name implies, is a commodity that arises incidentally but unavoidably, in the production of something else. If no use can be found for the byproduct it has to be discarded, and if it is noxious will cause pollution. A stern moralist may condemn any discard as waste, but it would be fairer to label a useless byproduct as an undeveloped resource – not every businessman is blessed with the ability to turn unconsidered trifles into gold. The preparation of cotton for spinning offers a telling example. The crop harvested from the cotton bush consists of seeds surrounded by cotton wool. In 1793 Eli Whitney invented a gin that simply and quickly separated the useful cotton wool from the unwanted seed. Nearly sixty years elapsed before men realised that the seed had its uses too: when crushed it yielded oil for soap-making, and the residue – oilcake – could be fed to cattle. The troublesome 'wasted' seed had become the raw material of a new industry.

A commodity with a useful byproduct (say mutton with wool) constitutes what economists call joint supply. Early economists were not much interested in the problems of production and contented themselves with describing pin-making, a handicraft that showed the virtues of division of labour. It was left to the mathematician

Charles Babbage to notice the use of 'materials of little value': some of these were examples of scrap recycled; others – goldbeaters' skins made from the offal of animals, and prussiate of potash from hoof and horn – were genuine byproducts though Babbage did not use the term itself.* Byproducts arise when complex materials, often but not always of organic origin, are broken down into their constituent parts, or recombine in a new way. A slaughterhouse gives rise to many profitable byproducts as the great meat-packing firms of Chicago demonstrated long ago. The coking of coal and the refining of oil are processes rich in byproducts. The alkali trade in Victorian Britain was long notorious for its failure to collect and use the harmful substances left behind in the making of soda. Even the slag left over in huge quantities after the smelting of iron ore had its uses. There is not much doubt that a long book would be needed to do justice to the economic history of byproducts. The purpose of this chapter is more modest and less indigestible – to give some impression of the opportunities taken or spurned by some British industries over the past two hundred years. It is an untidy story, a mixture of success and failure, with no clearcut trend towards or away from full use of the byproducts available.[1]

BYPRODUCTS FROM COAL

The destructive distillation of coal provides a spectacular starting point for an enquiry into byproducts. The coking of coal probably began in the middle of the seventeenth century. The process closely resembled the making of charcoal from wood: in both processes, volatile materials were driven off and something close to pure carbon remained; the use of coke or charcoal in industry removed the risk of contaminating iron, malt, or any other delicate substances with the noxious residues present in coal and wood. But unless the residues were collected and used, the manufacturer paid a heavy price for a purer fuel: a ton of coal will yield from 12 to 16 hundredweight of coke, and the coke has a lower calorific value than the coal from which it is derived. Neglect of the residues or byproducts was therefore a serious loss. For more than a hundred

* Byproduct is first recorded by the *OED* in 1857. It does not appear in J S Mill's *Principles* (1848) but was used by A Marshall, *Principles of economics* (1890) as if it were a familiar term not needing explanation.

years after coking began, nobody succeeded in making profitable use of the byproducts, and few even tried. The coke was reward enough, the byproducts an unfortunate nuisance.

Archibald Cochrane, 9th Earl of Dundonald, in 1781 secured a patent for the production of tar by distillation from coal. He was the first inventor to have even modest success in the business of winning byproducts from coal, but he had chosen a difficult path – and it was not long before a simpler way forward was found. Instead of exploring the complex chemistry of coal tar, inventors and businessmen found it easier and more profitable to use the gases given off by coal for gas-lighting. Indeed, for the gas companies that sprang up early in the nineteenth century and for the municipal gasworks established somewhat later, gas was the main product derived from the coking of coal. Coke itself was the chief byproduct and of course readily saleable. Problems arose only with the tarry and ammoniacal residues. Untreated they could be burnt, supplying some of the heat needed for coking the coal, but without any refining the residues had little value. The tar was rather thin for caulking ships' timbers, and would not have been a satisfactory ingredient of tar-macadam – which was in any case little used before the twentieth century.

It was many years before gas companies took their byproducts seriously. Although the price of gas tumbled from a typical 15s per 1,000 cu ft in the 1820s to 10s in the 1830s and to about 5s in the 1860s, increases in efficiency allowed gas companies to remain profitable. And their attitude to byproducts (apart from coke) was correspondingly cavalier. In 1866 Pontefract gasworks was getting a good price for its tar, which fetched a guinea a ton, but it let its ammoniacal liquor go to a chemical works at only 5s a ton; local farmers gave no more than 1s 6d a ton for lime that had been used to extract impurities, especially sulphur, from the gas. Castleford gasworks had a simpler solution – it sold all its tar for £5 a year, and its ammoniacal liquor for little more!

Gas prices continued to fall until the Second World War: in London the charge for a thousand cubic feet of gas was at most 2s. 6d. in 1912 and only 1s. 8d. in 1939. In these circumstances the expansion of the chemical industry did much to maintain the profitability of gasworks. The discovery of aniline dyes in 1856 gave a temporary fillip to the demand for tar, but the British dyestuffs industry made little headway until German competition ceased in the First World War. Tar distillers therefore dropped their prices once more: Sunderland gas company sold its tar to London ship-builders in 1849 for 2d. a gallon; after the collapse of aniline demand, the tar sold for only a ½d. a gallon, and had recovered to only 1¼d. by 1899, although the price of coal to the Sunderland gasworks had by then quadrupled – from 3s. to 12s. a ton. What the company lost on tar it gained on ammoniacal liquor, which sold at 3s. per 1000 gallons in 1849, and at 40s. fifty years later. The residuals including coke covered 65 per cent of the cost of coal in 1849 and 70 per cent in 1899. The low price of tar in Sunderland in 1899 may have been partly due to a combination among tar distillers. There is good evidence of monopolistic combinations elsewhere in the country, and gas companies and municipal corporations complained of the low prices offered by the confederate tar distillers. For example, in 1892 tenderers offered only eight to nine shillings a ton for Manchester's tar; after negotiations the corporation secured 14s., but later struck a deal with a maverick distiller, J E C Lord, who was willing to pay 21s. 9d. The South Metropolitan Gas Company, offered similar terms by Lord, chose to set up its own distillery.

Several other large gas undertakings also sought to diversify into the chemical industry through working up their own residuals. Even more alarming to chemical firms was the gasworks' growing practice of buying in supplies from nearby smaller undertakings, which had previously had to sell to manufacturing chemists. Without going so far as to make aniline dyes, a gas company could market a range of products from tar: pitch for roadmaking; creosote for timber treatment; crude naphtha from which light oils could be extracted; and anthracene, a source of dyestuffs and candles. It could also make the fertiliser ammonium sulphate: ammonia was readily distilled from the watery fractions of the tarry liquor, and sulphuric acid was derived from the process of gas purification. Without purification, coal gas stank of rotten eggs from the presence of sulphuretted hydrogen: the sulphur in coal was mostly driven off in making coke, and combined with some of the hydrogen that was a constituent of coal gas. The earliest treatment of coal gas took the form of passing it through lime – hence the evil-smelling fertiliser sold cheaply to farmers. A later and better way of extracting sulphur was to pass coal gas through a bed of hydrated iron oxide. The iron oxide took up the sulphur to form iron sulphate or spent oxide. It was then a simple matter for a manufacturing chemist to make sulphuric acid, restoring the spent oxide to its original condition and returning it to the gasworks for another tour of duty.*

* Note that spent oxide is a byproduct. It is also recycled.

Enterprising gas companies and a few municipal undertakings – Manchester and Coventry – saw no reason why they should not make sulphuric acid themselves, and use it to produce ammonium sulphate. Chemical firms protested at this encroachment on their business, but got little sympathy from the select committee that investigated the matter in 1912. There was a substantial output to contend for, though a diminishing proportion of the whole make of ammonium sulphate. In the nineteenth century, when beehive ovens produced most of the metallurgical coke, the competition with gasworks ammonia came from the distillation of shale oil, mostly in Scotland. As beehive ovens gave way to byproduct ovens at collieries and ironworks, so the supply of ammonia increased. After the First World War the production of synthetic ammonia grew rapidly and by the 1930s byproduct ovens as well as gasworks were finding it difficult to compete. For the gas industry the problem ultimately solved itself with the switch to natural gas and the closure of gasworks using coal. Coke ovens found themselves with unwanted ammonia and with surplus coal gas that the gas industry no longer wanted. Natural gas has a higher calorific value than coal gas and once burners had been adapted to use the new fuel, coal gas was incompatible. Unwanted gas had to be flared, and ammonia burnt under stringent safeguards against pollution. The alkali inspectors were helpless and saddened spectators of this apparently unavoidable waste.

Table 10.1 Production of ammonium sulphate

	(*000 tons*)	
	Derived from gasworks	*Total*
1862	5	5
1886	82	106
1899	137	208
1907	105	264
1912	179	396
1924	151	362*
1968	nil	862**

Sources: Census of Production 1907; Reports of alkali inspectors; Lyon Playfair provided the 1862 estimate for the noxious vapours committee.
*Includes synthetic.
**Production of ammonia.

Gasworks had fewer problems with the disposal of spent oxide. It was cheap and easy to recover sulphur from it for making sulphuric acid, and for as long as there was coal gas to purify there was a ready sale for spent oxide. The production of gas grew rapidly in the nineteenth century. In 1882, the first year for which comprehensive statistics are available, output reached 0.333 billion therms; output trebled by 1912 and trebled again between 1912 and 1955 (see Table 10.2). Until early in the twentieth century nearly all the gas made was coal gas.

Table 10.2 Output of gas 1882–1987

(*billion therms*)			
1882	0.333	1955	2.901
1912	0.995	1964–5	3.219
1920	1.177	1972–3	10.200
1940	1.443	1981–2	16.900
1948	2.119	1987	19.400

Sources: Gas Council, *Reports* 1951, 1965; *AAS*

From then on, supplies of coal gas were supplemented with water-gas and producer-gas, and from the 1960s with refinery gas and natural gas. Production of coal gas peaked in the years 1954–6 at 2.1 billion therms, and then declined until the last gasworks closed in 1976. There was a high correlation between the output of coal gas and the supply of spent oxide, and until the second world war gasworks were the main local source of sulphur for making sulphuric acid. After 1945 the output of anhydrite (calcium sulphate) from ICI's Billingham mine grew rapidly until by 1955 it dwarfed the contribution of spent oxide to sulphur supplies. Larger supplies of pure sulphur from overseas mines and from UK oil refineries in turn made anhydrite redundant, and the mine closed in 1975.[2] The deposits constitute a useful reserve of sulphur that may be called upon at some future date. In this respect they resemble Britain's increasingly neglected coal reserves – temporarily unwanted treasures.

ALKALI MANUFACTURE

Alkali manufacture like the gas industry took a long time to find a profitable use for its byproducts, which were quite as noxious as tar and ammonia. The Leblanc process for the manufacture of soda threw up two troublesome byproducts – hydrochloric acid and calcium sulphide. It will be remembered that the Alkali Act 1863 required that 95 per cent of the hydrochloric acid gas should be condensed, but without making any provision for the safe disposal of the acid when condensed. Chemists had long known that chlorine and some of its derivatives made valuable bleaching agents, and Tennants the Glasgow chemical manufacturers were making bleaching powder as early as 1799. The demand from the textile industries remained disappointingly small however, for many years. For example, it appeared that in 1852, on the basis of returns from four-fifths of the alkali trade only 13,000 tons of bleaching powder and 6,000 tons of bicarbonate of soda were made. Both these chemicals helped to use up hydrochloric acid generated in the making of soda – but only a little of it, for there were over 80,000 tons of chlorine in the salt decomposed in the alkali manufacture that year. The quantity of salt used doubled in the next ten years, and doubled again by the early 1880s. According to David Gamble, a prominent alkali manufacturer, the production of bleaching powder and chlorate of potash was fast catching up with the output of hydrochloric acid, at least in St Helen's. But his statistics claimed no more than that 18,000 tons of powder and chlorate were being made in 1876 – a quadrupling of output in fourteen years – and the smell of rotten eggs in south-west Lancashire contradicted his claims all too convincingly. The price of bleaching powder fluctuated wildly: it was sometimes more than £20 a ton, sometimes as little as £7. Naturally, less acid ran to waste when the price of bleaching powder was high.

In theory when less acid was wanted for bleaching powder more could have been used to treat the other byproduct, calcium sulphide. There was no lack of processes for recovering sulphur from calcium sulphide, but it was one thing to devise a process, quite another to persuade a large number of manufacturers to adopt it. Ludwig Mond, then a young German chemist employed in the Widnes alkali industry, patented one such process in 1862. At best it recovered only half the sulphur present in alkali waste; it sometimes left a residue that smelt as badly as the untreated waste; and it was a process that needed a good deal of land. By the time Mond's patent expired fourteen years later only four English alkali manufacturers were working it, and for all practical purposes it was a failure. The Chance-Claus process worked out in the years 1882–7 had much greater success. It recovered most of the sulphur in alkali waste, leaving a harmless residue. The discovery was timely, for by the late 1880s 0.75 million tons of waste were generated every year containing about a hundred thousand tons of sulphur. Firms working the Leblanc process found that they could usefully supplement their profits by recovering sulphur on the new plan, and at the same time remove an important source of atmospheric pollution. In the end the elaborate Leblanc process could not be saved by sulphur recovery. A simpler way of making soda – the ammonia-soda or Solvay process – rapidly displaced the older method in Europe, and more slowly in Britain. It used carbonic acid and ammonia – a byproduct of gasworks and of byproduct coke ovens – instead of sulphuric acid for reaction with salt; the soda was cheaper, there were no unpleasant byproducts like sulphuretted hydrogen, and means were soon found to recover chlorine from the ammonium chloride left after the reaction was complete. In the beginning Ludwig Mond, who with John Brunner pioneered the Solvay process in England, did not foresee its long-run competitive advantage over the Leblanc process. At first Solvay was a one-product process, soda, because chlorine could not then be recovered for conversion to bleaching powder. But within a few years that problem was solved and the success of Brunner-Mond assured.[3]

SOAP-MAKING

Soap-makers were among the most important customers of the alkali trade. Soap is made by combining oil or fat with alkali to produce a milder cleansing agent than alkali on its own. To a large extent the oils and fats supplied to soap-makers were byproducts from other trades. It is a curiosity of industrial history that alkali manufactured to the accompaniment of troublesome byproducts should itself be combined with other byproducts to make soap – which in turn generated another byproduct, glycerine. Traditionally soap-makers had competed with candle-makers for tallow from slaughterhouses and flushings (waste fat) from tanneries. Tallow made hard soap; for soft soap oil was needed. Whale oil and perhaps fish oil could be regarded as principal products rather than byproducts. But new sources of oil in the nineteenth century were in part at least genuine byproducts. Cotton-seed oil has already been mentioned, and oil from cocoa waste was also used in soap-making. The Yorkshire woollen and worsted industry for many years neglected some valuable byproducts, but eventually came to recognise their worth. These byproducts arose in the washing of wool and cloth. Unwashed wool contained large quantities of grease and sweat or suint, which was rich in potash: a stone of 16 lb of greasy merino wool might yield 5 lb of suint in washing; English wools commonly lost 3 lb per stone in washing. It appears that until well into the nineteenth century the scourings of wool ran into the rivers without any attempt to recover the grease and potash. The grease after purification could be turned into soap or candles; the potash was a valuable fertiliser.* Woollen yarn had to be oiled before weaving, and it was therefore necessary to wash the cloth in soft soap to remove the oil after manufacture. In a mill conducted with due economy the owner would have arranged for the soap suds and oil to be collected and returned for recycling into soap and cloth oil, but as with wool scouring, river pollution was often the line of least resistance.

For small firms especially, the profits to be earned from attention to byproducts were trivial. When recovery was attempted it was often outside contractors who conducted the business. Wool scourings and soap suds were too bulky to be collected and removed without some preliminary treatment, and it became the practice for millowners to provide heat and power at the mill. This enabled the byproducts to be filtered, evaporated and pressed before removal for further refinement. William Teall, a prominent grease extractor in Wakefield, set up in business in 1852. It seems unlikely that there or elsewhere much serious attention had been paid to recovering oil before the mid-1840s. By 1866 Teall was recovering every week 50 tons of clear fatty matter containing cloth oil, besides a bulky residue for sale to manure (fertiliser) makers. Only the finest olive oil was used for oiling yarn and at £60 a ton it was not cheap. Recovered oil was worth about £20 a ton, yet careless millowners allowed the grease extractors to pay derisory sums for the right to treat soap suds and oil. One manufacturer, David Yewdall, received only £50 for suds and oil that had cost him £5,000. It was estimated that the material used in scouring wool and washing cloth cost £1.2 million a year, of which a mere £100,000 was recovered in the 1860s. The royal commission on rivers pollution argued, with expert support from grease extractors, that at least twice that amount could be recovered – with benefit to

* Pig's dung and urine, and human urine, for which a firm's employees could charge 5s a barrel, were favoured scouring agents, and fertilisers in their own rights, if collected.

manufacturers as well as to the rivers of the West Riding. The industry was slow to learn, for Bradford's sewage works were still conducting a profitable trade in grease after the Second World War, and in 1946–7 sold materials worth £128,000 for greasing railway axles, and making powdered soap and paint.

The Swedish chemist Scheele discovered in the eighteenth century that glycerine was a constituent of all fats and oils. In soap-making, caustic soda combined with the fatty acid in oil or fat, taking the place of glycerine, which was left behind as a byproduct. Glycerine also arose in bone-boiling and in candle-making. As with many byproducts, at first it had few uses, but by the 1870s it was valued in pharmacy, brewing, confectionery, perfumes and explosives. The well-known firm Price's Candles was producing glycerine as a sideline to its hard white candles by 1840. When pure it commanded a high price but when it occurred in soap-making it was far from pure. William Gossage was a soap-maker famous for his 'Gossage tower' for condensing hydrochloric acid gas, and a man eager to use science in the service of industry. Yet in the 1860s his firm was letting five tons of glycerine run to waste in the Mersey every week. If pure like the glycerine arising in the making of hard candles, it would have been worth 1s. 4d. a pound or £150 a ton; undistilled it was probably worth only half as much, money that Gossage's was apparently prepared to despise. At the time of the first census of production in 1907 glycerine was worth less than £50 a ton, thanks in part no doubt to greater care in recovering it in soap-making. Output then stood at 17,000 tons, little more than two per cent by weight of the combined output of soap and candles. Full recovery would have more than doubled the output of glycerine – and unless demand was elastic perhaps halved its price. Profit and conservation make uneasy companions on the road to efficient use of resources.[4]

FERTILISERS

Farmers in nineteenth-century Britain mostly kept their land in good heart by rotation of crops and the use of farmyard manure. There was only a rudimentary fertiliser industry, so rudimentary that chemical fertilisers were usually called by the old-fashioned and misleading name of manure. The usage at least had the virtue of reminding the public that fertilisers, like manure, were largely derived from byproducts, or from recycled waste.* This was obviously the case with the dust from rag-grinding in the shoddy trade, with waste wool mixed with workers' excrement, and with the road sweepings of London and of Covent Garden's fruit and vegetable market. Nowadays such rubbish fills gravel pits and exhausted brickfields. In the nineteenth century farmers were willing to pay a little for large quantities of it. A thousand loads of mud and horse manure were removed from the streets of the City of London every week. As a concession to farmers the River Lee Trust carried manure free of charge from London into Hertfordshire, but even this generous offer attracted relatively little business: in 1865 the 24,000 tons of manure carried up the Lee in barges accounted for only six-and-a-half per cent of the total traffic.

A less obvious step to promote higher fertility was the use of crushed bones, a practice known in the eighteenth century and coming into wider use in the 1820s and 1830s. Imperfectly crushed bones would release phosphates only slowly into the soil. Chemical treatment was needed to give farmers readier access to the nutrients they sought. It was J B Lawes, the founder of Rothamsted experimental farm, who in 1843 introduced the manufacture of superphosphates at his Deptford factory by dissolving bones in sulphuric acid. It is impossible to know the quantity of bones collected in Britain by itinerant rag-and-bone men and by marine-store dealers. One estimate, made in 1904 when the trade was probably in decline, suggested a supply of 60,000 tons a year. The import trade in bones is more reliably recorded: it reached 27,000 tons in 1850 and was at its height between the 1860s and the First World War. Not all bones of course were converted into superphosphates: some were used for glue-making, a few were burnt to make 'animal charcoal' used in sugar refining and shoe polish.

* This was true even of ammonium sulphate, the ammonia being a coal-tar byproduct. However, there was little demand for ammonium sulphate in Britain before the First World War, and seven-eigths of output was exported.

Table 10.3 Average annual imports of bones

	(000 tons)
1861–9	69
1870–9	87
1880–9	65
1890–9	68
1900–9	49

Source: *Agricultural Returns* v.d.

Bones were richer in phosphorus than were mineral sources, but the lower price of rock phosphates allowed them to take an increasing share of the market. The first mineral supply to be exploited was a deposit of coprolites quarried in East Anglia. At peak output in 1876 the deposit yielded a quarter of a million tons of phosphatic rock, but the source was quickly exhausted and production had virtually ended by 1900. A more persistent competitor to bones was rock phosphorus from the United States. It is not clear either from British or from American sources when this trade began, but it seems to have been in existence on a small scale by the early 1870s, and by the end of the century was worth twice as much as the import trade in bones, and was six times as bulky. The import of rock phosphorus continued to grow and in the 1950s exceeded a million tons a year, mostly coming by then from North Africa.

The mineral supply had dwarfed the organic byproduct supply from bones. Another organic source, the Peruvian guano deposits, was first exploited in the 1830s. Since guano consisted of bird droppings accumulated and compacted over time it was not to be expected that the deposits would have a long life. They were owned by the government of Peru and in the 1870s Peruvian bondholders in Britain as well as farmers had an interest in knowing how long the guano would last. The Royal Navy dispatched an expedition to enquire into the matter. After some delay Peru gave permission for boring to take place. The results were not encouraging. The guano deposits varied in depth and density and were in places obscured by sand. The subsequent course of trade showed that guano provided only a temporary supply of phosphates for a hungry world.

Basic slag, a byproduct from the Bessemer converter and open-hearth furnace, was to provide a more persistent supply of phosphates. Phosphoric iron ores, from which basic slag was derived, were of little use in steel-making until the advent of the Gilchrist-Thomas process in 1879. In 1885 field experiments with ground slag – 'basic cinders' as it was then known – demonstrated its worth at the Durham farm of the Carlton Iron Works Company. The most telling results came from Cockle Park, the experimental farm run by Northumberland County Council. In 1897 and after, William Somerville, the Director of Cockle Park, showed that a simple dressing with cheap basic slag gave better grass yields than more elaborate and expensive treatments with other fertilisers. It later turned out that under different conditions in other parts of the country the use of basic slag was not so effective. Even so, a once useless waste product could now be sold, after grinding, as a valuable source of phosphorus. By 1904 output of ground slag had reached 300,000 tons a year, of which half was exported. Enthusiasm among iron-masters seems to have waned in the 1930s: in 1935 when half of all steel was made from phosphoric ores the production of ground slag amounted to no more than half a million tons.[5] It is probable that much more could have been produced if farmers had been willing to buy it. The ores from which basic slag was derived would not have been worth mining for their phosphorus alone.

Other rocks richer in phosphorus were available. They (and deposits of potash) were extensive but far from inexhaustible. When Sir William Crookes gave his famous address to the British Association for the Advancement of Science in 1898 he was concerned about a prospective shortage of nitrogenous fertilisers rather than of phosphates and potash. The perspective later changed. Almost unlimited quantities of synthetic ammonia were made by fixing atmospheric nitrogen, using the Haber process and much expensive equipment; and such supplies were assured in Britain within about thirty years of Sir William's address. The outlook for mineral phosphates and potash is much less certain, though farmers have no immediate cause for alarm. At present their feeling is, rather, one of embarrassment at the huge quantities of manure generated under conditions of intensive farming. Most farmers are businessmen first, and conservationists a very bad second: to them organic farming is a cause with a distinguished patron – and very few followers outside the towns. Not content with the nutrients supplied by farmyard manure and careful rotation of crops, farmers apply to the land large and easily administered doses of phosphates and potash derived from exhaustible mineral deposits. They face the future in the same negligent spirit as Louis XV.

BUILDING MATERIALS

The demand for slag in agriculture did little to reduce the huge quantities of slag, acid and basic, arising in the smelting of iron ore and the making of steel. Slag from the Middlesborough ironworks was being dumped at sea in the nineteenth and early twentieth centuries. But on the opposite north bank of the Tees a more constructive use was found for it. Under the Tees Conservancy Act 1858 large areas of the foreshore were reclaimed by controlled dumping of virtually indestructible slag. By 1911 nearly seven square miles of land had been or were being reclaimed for industrial purposes. When the reclaimed land was sold, half the proceeds went to the Tees Conservancy, a quarter to the Crown and a quarter to the frontagers. Sir Hugh Bell, proprietor of the Clarence Works on the north bank of the Tees, was naturally an enthusiastic tipper of slag. Blast-furnace slag also makes good hardcore for road-making, which is just as well since this kind of slag makes up over half the waste arising in a steelworks. A plant producing three million tons of steel will generate 600,000 tons of blast-furnace slag; much of the steel slag will be recycled in the blast furnace, or will be ground into fertiliser, and only about a quarter of the waste in a modern plant finds its way to the works tip.

Other waste was also likely to find itself usefully employed in construction. Every works driven by steampower accumulated a heap of ashes of great bulk and little value. Fine ashes were sometimes mixed with excrement and sold as manure for trivial sums to local farmers. Or they were mixed with more usual ingredients, sand and lime, to make mortar. Coarse ashes – clinker and no doubt some cinders too – made a cheap roadbed when new railways were being built. Edwin Brooke, a Huddersfield manufacturer of drainpipes, had accumulated 20,000 tons of ashes at his works in 1866. He was steeling himself to grind them up and mix them with clay for brick-making; much to his relief a railway company solved his problem by taking them off his hands for ballast, that is to make the railway bed. Ashes also found a use filling potholes in the road – with the disadvantage that heavy rain tended to wash the finer materials away into drains and rivers. In more recent times the electricity industry has found a market for part of the huge quantity of ash it produces every year. The ash from powdered fuel (PFA, or pulverised-fuel ash) is suitable for making cement and lightweight insulating bricks. When an industry is producing 9 million tons of fine ashes a year it is some relief to sell a third of them to building and civil engineering firms. The anhydrite much used for about fifty years in the production of sulphuric acid had chalk as a byproduct. This, like ash, could readily be used in making cement, or the compound fertiliser nitrochalk. The china-clay industry is another that generates a great deal of bulky waste, for only part of the clay washed from the pits is fine enough for use in pottery, paper-making, pharmaceuticals and many other industries. For many years the coarser clay was left behind in huge white spoil heaps creating a landscape of eerie desolation. Recently the industry has had some success in finding a market for its waste among makers of bricks and cement.

THE DECLINE OF BYPRODUCTS?

To byproducts there seems to be no end, but it may be that they were relatively more common in pre-industrial economies than they are now. Before industrialisation, economies relied for many raw materials and intermediate goods on natural products derived from living organisms. Timber, wool, cotton, leather, dyestuffs, and animal and vegetable oils spring readily to mind. Among intermediate goods, ammonia was derived from horn, urine, or seaweed; ink was a mixture of lampblack and linseed oil; gunpowder a mixture of saltpetre from stables with charcoal and sulphur: only the sulphur came from a mineral, not an organic, source. Inventions, scientific discovery and the growth of demand induced a shift in sources of supply to minerals and more recently to synthetic materials. By synthetics we mean materials made with the aid of science and advanced technology: ammonia made with atmospheric nitrogen, artificial silk, nylon, and plastics are examples of products fitting the synthetic description. There is no economic or scientific law that decrees an organic – mineral – synthetic cycle, only a plausible generalisation or an assertion of likelihood. Building materials have passed through all three phases: timber; stone and bricks; steel, glass and plastics. Textiles and to a large extent dyestuffs have skipped the mineral phase, though synthetic fibres and dyes are of course derived from minerals – oil and coal – and these minerals are largely, if not exclusively, of organic origin. Phosphoric and potassium fertilisers have stopped short of the synthetic phase unless the simple treatment of bones and rock with sulphuric acid can be described as high technology.

Nitrogenous fertilisers, in contrast, have passed through the full cycle: farmyard manure and crop rotation represented the organic phase; the alkali industry and coal distillation for ammonia depended on mineral resources aided by a little chemistry; the fixing of atmospheric nitrogen was a triumph of high technology. The discovery of new mineral deposits sometimes puts the cycle into reverse. With simple techniques sulphur was once obtained by roasting pyrites as an alternative to mining in Sicily; it was later recovered from alkali waste and extracted from anhydrite or from crude oil with the aid of much capital and scientific skill. Now, new mineral deposits have made the use of anhydrite, though not of oil, obsolete, so that there has been a partial return to the methods of low technology.

Low technology, and especially reliance on organic and therefore chemically complex resources, throws up many byproducts.These were in Alfred Marshall's mind when he was discussing economies of scale:

> The chief advantages of production on a large scale are economy of skill, economy of machinery and economy of materials: but the last of these is rapidly losing importance relatively to the other two. It is true that an isolated workman often throws away a number of small things which would have been collected and turned to good account in a factory; but waste of this kind can scarcely occur in a localised manufacture even if it is in the hands of small men; and there is not very much of it in any branch of industry in modern England, except agriculture and domestic cooking. (Marshall, *Principles of economics*, pp 278–9.)

Marshall deceived himself in supposing that waste of materials could scarcely occur in industry even when conducted by small men. Every generation masters some problems of waste recovery or disposal, only to create others for succeeding generations to tackle. The history of pollution and of the waste trades generally, shows that Marshall's complacency is and is always likely to be unjustified. It is much harder to decide whether 'economy of materials' has become less important in modern times. On the one hand poor economies cannot afford to waste any resource that can be put to use. On the other, only rich economies with large-scale methods of production accumulate enough 'waste' (say, grease from wool-scouring) to make recovery worthwhile.

Evidence would be preferable to argument, but reliable evidence is hard to come by. Since 1801 there has been a ten-yearly population census. It has been organised with a variety of users in mind, but it is pretty clear that the historian of waste and byproducts has not been among them. The earliest census to attempt a detailed description of occupations was that of 1831, and the attempt was confined to men aged twenty and over in handicrafts and retail trade, that is, only 30 per cent of adult men were included. Few of the occupations had any obvious connection with waste trades: a handful of bone boilers and dealers in marine stores; some nightmen and scavengers; soot and chimney sweeps; and most numerous of all, old clothes dealers and rag dealers. Even in Middlesex, where much of the trade in old clothes and rags was to be found, only one per cent of those engaged in retail trade or handicrafts apparently earned their living in the waste trades. The 1841 census tried to describe the occupations of the whole workforce, and detected dealers in metals and paper as well as the rag dealers and others noted ten years earlier. Nevertheless, the number apparently employed in waste trades remained below 3,000 and was indeed slightly smaller than in 1831. In 1851 the number of waste trade dealers rose above 5,000, mostly because of a jump from 471 to 2,068 persons described or describing themselves as marine-store dealers. The number of rag dealers doubled in the next fifty years, and suspiciously doubled again between 1901 and 1911. The 1981 census is no more helpful than the others, and at best supplies a total of about 4,000 persons engaged in the scrap and waste trades. A private census conducted by the British Reclamation Industries Confederation in 1978 uncovered a much larger number. Returns from the 900 member firms showed that they had 20,000 employees. In addition there were 10,000 dealers registered under the Scrap Metal Dealers Act 1964, few of whom belonged to the BRIC. Though mostly in a small way of business these metal dealers usually employed a few men and boys. A conservative estimate therefore suggests that some 50,000 persons had jobs in the private sector collecting and handling scrap and waste, not the mere tenth of that number reported in the census of 1981. In addition there were the mostly non-profit-making jobs in the collection and disposal of refuse by local authorities; and some employment in manufacturing industry recovering and using scrap and recycled material. Any estimate of the total numbers employed in the scrap and waste trades is bound to be rough and shaky. A figure of 300,000 plus or minus 100,000 is put forward only with due hesitation. If this estimate is near the truth it suggests that about one per cent of the workforce is employed in the waste trades. This is a small proportion of the gainfully occupied, but equivalent (at the upper end of the range) to employment in agriculture or the chemical industry, or (at the lower end) to employment on the railways.[6]

Censuses of production throw as little light on the place of waste trades in the economy as do the censuses of population. In 1907 when the first census of production was taken the statisticians noted sales of byproducts, offals, waste products, spent grain, sprouted barley and many more. They did not take account of waste or scrap reused within a works, because that was a transaction internal to a firm: no money changed hands and in most cases no value would have been imputed to the scrap material. To that extent the census understated the importance of recycling. On the other hand, where sale did take place it would be easy to exaggerate the size of the waste trades. It would be extravagant, for example, to class sales of coke by a gasworks as a waste trade: nobody had ever neglected or thrown away coke in the way that tar and ammoniacal liquor had run to waste or been sold at give-away prices. Yet coke was a byproduct of the gasworks just like the other residuals. It is unfortunate and confusing that the term waste has overtones that do not attach themselves to prosaic words like byproduct, offal, and residual. From the census of production no very clear picture emerges. At one extreme the offals and other waste products of distilleries amounted to ten per cent of the gross output; and if yeast, a still more valuable byproduct, was included the proportion rose to 25 per cent. At the other end of the scale, in the production of newspapers, periodicals and magazines, waste products amounted to only one per cent of gross output. Moreover, since it was not easy to say where ordinary production ended and waste products began, different firms and observers might have drawn the line in different places. An average figure for waste products is therefore hard to arrive at, but it seems likely that in 1907 one per cent would be too low and ten per cent much too high.[7] Later censuses, for example those of 1935, 1968 and 1973, still show a substantial output of 'waste' products. There is no reason to think that they have ceased to be important in the past twenty years, though recent production censuses are less than informative on the point. Marshall's conjecture that economy of materials contributed less to welfare with the passing of time was probably unsound when he made it, and has had little validity since, least of all now that there is widespread concern about the depletion of natural resources.

REFERENCES

1. P L Simmonds, *Waste products and undeveloped substances* (1873), pp 196–207; C Babbage, *Economy of machinery* (1832), p 11.
2. M E Falkus, 'The British gas industry before 1850' (*EcHR* 1967), pp 501–2; r.c. on rivers pollution, *Third report* (PP 1867 XXXIII), qq 12537 ff, 15309–33, 15829– 40; *Second report* (ibid.), qq 2925 ff; *36th report on alkali works* (PP 1900 X); s.c. on gas undertakings (residual products) (PP 1912–13 VII), *Evidence*, qq 1434, 1611, 1682–6, 2328 ff, *Report*, pp 1–2; R A Mott, 'The domestic smoke problem: the possibilities of coke–oven fuel' (NSAS, *Proceedings of the fourth annual conference 1932*), p 25; A O'Connor, 'The prevention of pollution from industry: the steel industry' (NSCA, *Clean air conference 1974*), p 4; *Industrial air pollution 1975* [formerly alkali inspector's reports], p 16; *105th report on alkali works* (1969), pp 15, 42; *111th report* (1975), pp 20–2, 78; *AAS* v.d.
3. Lords committee on noxious vapours (PP 1862 XIV), qu 1507; r.c. on noxious vapours (PP 1878 XLIV), qq 4753, 4764, 4943 ff, 4966, 4979, 5029; D W F Hardie, *Chemical industry in Widnes* (1950), pp 126–8; *25th report on alkali works* (PP 1899 XVIII), pp 11–13; *29th report* (PP 1893–4 XVI), p 14.
4. R Campbell, *The London tradesman* (1747, repr. 1969), p 263; H J Spooner, *Wealth from waste* (1918), p 205; r.c. on rivers pollution *3rd report* (PP 1867 XXXIII), qq 14505–6, *report* (ibid.), pp xxiii-xxxv; ibid. qq 7200 ff; *Municipal yearbook for 1949*, pp 1176–7; *Victoria County History: Surrey*, II ii, pp 405–7; r.c. on rivers pollution, *1st report* (PP 1870 XL), pp 36, 103.
5. Simmonds, *Waste products*, p 12; r.c. on rivers pollution, *2nd report* (PP 1867 XXXIII), App. p 161; s.c. on agricultural distress (PP 1836 VIII pt I), qq 5824, 5991, 7441–3; departmental committee on the fertilisers and feeding stuffs act 1893 (PP 1905 XX), qq 3203–9; *99th report on alkali works* (1962), pp 14–15; J R von Wagner, *Handbook of chemical technology*, (1872), p 553 ff; US Commerce and Navigation 1874, doc 1650; ibid. 1884, doc 2197; ibid. 1888, doc 2552; [UK] *Trade and navigation accounts* and *Annual statements of trade* v.d.; *Reports . . . relative to guano deposits in Peru* (PP 1874 LXVIII); J Wrightson, *The principles of agricultural practice* (1893), pp 176–8; J A S Watson and M E Hobbes, *Great farmers*, (1951), pp 135–7.
6. R.c. on coast erosion, *Third report* (PP 1911 XIV), p 126; *Evidence* (PP 1909 XIV), qq 14383 ff; A O'Connor, 'Pollution . . . the steel industry' (NSCA *1974*), pp 7, 12; r.c. on rivers pollution (PP 1867 XXXIII), qq 4056–61; ibid. (PP 1873 XXXVI), answers to queries, passim; Central Electricity Authority, *Annual report and accounts 1956–7* (PP 1956–7 XI), p 26; Central Electricity Generating Board, *Annual report and accounts 1965–6* (PP 1966–7 XXVIII), pp 13, 28; A Marshall, *Principles of economics* (7th edn 1916), pp 278–9, but substantially unchanged from the 1st edn 1890; Census 1841, *Occupational abstract* (PP 1844 XXVII), p 27 ff, 31 ff; *MRW*, 14 October 1978.
7. Census of production 1907, *Final report* (PP 1912–13 CIX).

CHAPTER ELEVEN

Conclusion: Economic Growth and the Environment

Although economic growth is not a term that the environmental historian uses with any affection, since he is dealing with the adverse consequences of growth, he can scarcely avoid it altogether. Any lessening of environmental evils contributes to psychic welfare, and reduction of waste adds to material income and promotes growth, other things being equal. An enquiry into waste and its prevention is therefore an enquiry into growth at one remove. So far, recycling and the use of byproducts have been considered as antidotes to waste of materials. But there are other ways of economising and they will be examined briefly in this chapter before concluding with some discussion of economic growth, past and prospective, true and false.

THE QUALITY OF GOODS

Before manufactured goods are produced, somebody – designer, engineer, salesman, an advisory panel of consumers – has to decide how well they are to be made. When a bridge or a piece of

Opposite:
The two Forth bridges, 1890 and 1964. The Forth railway bridge (1890) and the road bridge (1964), illustrate drastically different attitudes to problems of design and the use of materials. The powerful railway bridge and the slender road bridge pose important questions: where does economy of materials become cheese-paring, and where does solidity become extravagance? And which approach is more likely in the long run to promote economic growth and to conserve the world's resources?

machinery is under consideration it is necessary to allow for a safety factor. If the maximum weight a railway bridge is to carry is 5,000 tons, should it be designed to be safe when carrying two or eight times that weight? When a boiler is to operate at 200lb pressure psi should it be tested to 300lb or 1000lb pressure? Many manufacturers – of toys, cars, electrical goods, clothes, furnishing fabrics – have to take safety into account. Safety is just one of many problems of quality that producers have to consider. (Quality control is a different matter: the preservation of uniform standards of manufacture once a particular level of quality has been decided upon.) Decisions about safety and quality affect not only the care with which goods are made, but the amount of material that goes into them. For long-run economy of material, what is the right decision: skimp and scrap, or make goods that are sturdy and long-lasting?

Bridge-builders in antiquity had no doubt that they ought to make their bridges last. The Ponte Palatino built over the Tiber at Rome in 180 BC remained in use until 1598 when it was swept away in a flood. The Puente Trajan at Alcantara in Spain was built in the lifetime of the Emperor Trajan, who died in AD 117. Because it would have been useful to the enemy in the Peninsular War, it was partly blown up in 1809, but it is still recognisably a bridge. Even Roman engineers did not often build structures that would last so long and their modern successors almost certainly harbour no such ambitions. About Victorian engineers it is difficult to be sure. Their best work was solid and monumental: the Forth Bridge for example has completed its first century and shows no sign of old age. The Victorian engineer who built it made generous allowances for safety. William Fairbairn (1789–1874) used a safety factor of eight in some of his bridges and recommended the same factor in boiler-making, preferably with three safety valves. The usual safety factor in bridge-building appears to have been four or five.

These were the factors mentioned in the enquiries into the Tay Bridge disaster of 1879. The railway bridge across the estuary of the Tay opened in June 1878, providing a direct route south from Dundee. Eighteen months later on the night of 28 December 1879 the bridge fell during a severe storm, and a train was lost with its passengers and crew. Two enquiries followed: one was the usual Board of Trade enquiry into serious railway accidents; the other was conducted by a select committee hearing evidence about proposals for a new Tay bridge. Sir Thomas Bouch, the distinguished but elderly civil engineer who had designed the Tay Bridge, had, on paper, taken most of the precautions required by the best practice of the time. The piers were strong enough to take twenty times the greatest possible load they would ever have to bear. When Bouch discovered that for part of its length the bridge would rest on sand, not rock, the caissons were enlarged, but spaced further apart, and the piers above the waterline were built of iron rather than brick, because iron put less weight on the foundations. The enquiry reported that nothing had gone wrong with the foundations, that the girders and the permanent way were well made, and that the piers were adequate to support the vertical weight they had to bear. Yet there was a fatal weakness in the design – the engineer had made no special allowance for lateral wind pressure. At the time Britain had no official standard for resistance to wind. French and American engineers allowed for a force of 50lb per sq ft, and so too did William Fairbairn. If well-built the Tay Bridge could in fact have withstood a pressure of 70lb per square foot. Unhappily, there were serious faults of workmanship and supervision that quite offset the relatively small margin of safety. The clusters of cast-iron columns that formed the piers were of uneven thickness and poorly bolted together, and the cross-braces that should have given the structure rigidity broke away from the piers during the storm because of weaknesses in the cast-iron lugs to which they were bolted. As a result a third of the bridge – over 1,000 yards – fell, unable to resist the fury of the storm and the weight of the train.

The collapse of the Tay Bridge showed the uselessness of safety factors unsupported by good workmanship and close inspection. Disasters at sea, which used to be common and still occur too often, were sometimes caused by reckless seamanship. The loss of the *Titanic* in 1912 was a glaring example. Although popularly supposed to be unsinkable, the ship was designed to survive a collision if one or at most two of its water-tight compartments were torn open. When it struck an iceberg, four compartments were seriously breached and two slightly; the ship was therefore doomed. The officers had received and acknowledged warnings of ice, but the *Titanic* steamed on at her usual speed of 22 knots, and had much too little time to alter course when the iceberg was sighted. The Wreck Commissioner recorded a merciful verdict on the captain when he reported: 'What was a mistake in the case of the *Titanic* would without doubt be negligence in any similar case in the future.' (*Report on the loss of the . . .* Titanic, 1912–13, p 30.)[1]

The oil tanker *Torrey Canyon* came to grief in 1966 not because of any inherent fault in the ship, but because the captain took a short cut onto rocks off the Scilly Isles. Errors of seamanship apart, ships are often lost because naval architects, like bridge builders, are trying out daring new ideas. The unusual configuration of some classes of vessel makes it difficult to produce a wholly safe design. Car ferries and very large bulk-carriers have been particularly prone to accident: car ferries because they are top-heavy and have large bow and stern doors; bulk carriers because their great length exposes them to violent stresses.

It is difficult to guard against folly and human error, but knowledge of the properties of materials allows designers to estimate factors of safety with some confidence. Victorian engineers, perhaps less confident of the reliability of their materials, made extravagant allowances just to be on the safe side.* Some recent feats of bridge-building in Britain illustrate how far economy of materials can be taken when engineers are confident of their calculations and of the quality of their materials. The Forth road-bridge completed in 1964 was already slender by the standard of pre-war suspension bridges. With an overall length of nearly 6,000 feet and a main span of 3,300 feet, its chains weighed 7,400 tons and the total weight of steel amounted to 29,000 tons. The Severn road bridge completed two years later, with its main span only 60 feet shorter, saved 3,000 tons of steel in the chains and 10,000 tons overall: in total it was an eighth shorter but used only two-thirds of the steel needed for the Forth Bridge. It may be doubted if economy can be carried much further unless new materials come into use: within twenty years of opening, the Severn Bridge was causing some anxiety among civil engineers. The Humber Bridge opened in 1981 is equally slender and has a main span nearly half as long again as the Severn Bridge. Only time will show whether such bridges will last as long as the more solid works of the Victorian engineer.

Bridges cost so much that nobody can suppose that they will be designed to have a short life. Faults of design or workmanship may cause them to fall down early, but that is an unintended result. Other buildings and many goods are meant to have a short life. If the buyer knows and approves, or at least acquiesces, no harm is done. But if the maker produces an inferior article with intent to deceive, a charge of planned obsolescence can fairly be laid at his

---

* The generous safety factors in Victorian design probably owed something to cheap capital. At a time of rapid capital accumulation and low interest rates there was little point in cheese-paring.

door. Nineteenth-century cotton mills and farm buildings were solidly made and plainly meant to last; their owners evidently supposed that future technical changes could be accommodated within the existing buildings. Modern factories and farm buildings on the other hand are often cheaply and simply built and can easily be replaced if circumstances change. In company accounts freehold buildings are commonly depreciated at the rate of 2 or 2.5 per cent (that is, they have an expected life of 50 or 40 years), but some are reckoned to last only ten years. Tools, equipment, and computers naturally do not last so long as buildings: computers have an expected life of only five years, computer-controlled equipment and office equipment and furniture eight years, and fixed plant, machine tools and major equipment ten years. Accounting policies vary somewhat from company to company, and some equipment (telephone ducting, for example) has a comparatively long life. Since most accounts are presented on historical conventions, it is likely that firms cannot in an inflationary world replace their tools and equipment as often as rates of depreciation would suggest. Nevertheless the provisions made by companies presumably reflect their judgement of what would be prudent conduct in a stable economy.

The length of life of consumer goods is equally varied. Even soft goods like textiles can be designed for long life: Persian carpets easily outlive their owners. At the other extreme the cheapest clothing materials are worn out in months rather than years. 'These goods would look mere rags [admitted a Manchester merchant in 1831] without some filling in them, and you do not seem able to get a price for a better quality which would look decent without.' No doubt every generation of businessmen faces the problem of matching goods to the purse and taste of the consumer. In 1991, a high street jeweller made exactly the same point as the Manchester merchant, but in public, when he confessed that some of his wares were 'total crap'. He added in his defence:

> It is interesting, isn't it, that these shops, that everyone has a good laugh about, take more money per square foot than any other retailer in Europe. Why? Because we give the customers what they want. (*The Observer*, 28 April 1991.)

Fashion jewellery, it is clear, hardly qualifies as a durable consumer good, though few goods last as long as expensive and well-made jewellery. It is in this field of durable consumer goods that a suspicious public fears it is most likely to be deceived into buying an article that looks well, performs badly and dies young. Price is by no means an infallible guide to durability, as the motor car trade shows. On the reasonable assumption that the rate of depreciation of cars reflects their expected length of life, broadly speaking expensive cars* do last longer than cheap ones. But the correlation is by no means exact. Among a group of seven of the cheaper makes, mostly costing about £4,000 new in 1982, five had lost between 40 and 44 per cent of their value in four years, the sixth had lost 49 per cent; the worst performer, much dearer than the others, had lost 60 per cent. The larger and more expensive cars had somewhat lower rates of depreciation: one lost a mere 10 per cent of its value in four years, and three others lost between 29 and 39 per cent; two luxury cars, one British and one German, suffered a higher rate of depreciation than all but two of the cheaper cars. Of the thirteen models all but three lost between a third and a half of their value when new in four years. This is a moderate and reasonably uniform rate of depreciation, suggesting that manufacturers were trying to make their cars to standards not very different from those of their competitors. By the 1980s most makers were taking thorough anti-corrosion measures and offered guarantees against rust. As a result the average life of a car lengthened. Few could rival the Volvo, which according to the Swedish Consumers' Association lasted on average for seventeen years. But there can be no doubt that cars of the 1980s were in many respects superior to earlier models: they rusted less readily than cars made in the 1960s and 1970s; they were faster, had better brakes, used less petrol, had more precise steering and embodied safety features previously omitted. Whatever may be the case in America the cars sold in Europe now appear on the whole built to last a reasonable length of time.** This conclusion may surprise anybody who has seen the speed at which cars are assembled on the conveyor belt, or the demonic frenzy that possessed spot-welders in

---

* Data from Parker's *Car Price Guide*, May 1986. The cheaper cars: Polo, Astra, Metro, Cherry, Escort, Sunbeam and Lancia; the luxury cars: Mercedes, Bristol, Volvo, Rolls Royce, BMW and Jaguar. The second-hand prices were for cars rated A1, i.e. used cars in good clean condition with about 10,000 miles a year on the clock.

** The life expectancy of American cars in 1957 was only nine years (Battelle Memorial Institution, *A survey of iron and steel scrap*, 1957), quoted in D Gabor *et al.*, *Beyond the age of waste* (1978). According to Professor Gabor, more recent reports do not give reliable data.

the days before robots took their place. But there seems to be little evidence that nowadays manufacturers deliberately set out to make their cars last as short a time as possible.

To prove a charge of planned obsolescence would in any case be difficult because it implies conspiracy, a shadowy business in which the participants usually take care to cover their tracks. The term 'planned obsolescence' was popularised in Vance Packard's *The waste makers* (first English edition 1961). But the idea had occurred to suspicious minds earlier, and it is surprising that the *Oxford English Dictionary* has no example earlier than 1966. Whether planned obsolescence exists, or is only a nightmare of the imaginative, there can be no doubt that it is widely believed to occur. The following assertion (1970) is typical:

> Many of the mass-produced articles in use today have a built-in obsolescence and this process will accelerate in direct ratio to increasing national affluence. The effect is to produce a throw-away society engendering a formidable collection and disposal problem. (Higginson, 'Disposal of refuse . . . ' p 30.)

A disposal problem can arise because the consumer wishes to follow the latest fashion – whether in shoes, cars, or kitchens – and discards goods before they are worn out. It would be simple-minded to reduce the dictates of fashion to the level of a producers' conspiracy. Nor is it necessary to assume planned obsolescence if goods wear out quickly: poor quality in response to low purchasing power may be a more realistic as well as a more charitable explanation. Pessimists believe that quality deteriorates steadily as time goes by; it is more likely to fluctuate in accordance with the materials available, the taste of consumers, and the competitive pressures felt by producers. In the world of textiles the widespread substitution of cotton for linen, silk and wool during the nineteenth century undoubtedly cheapened goods and widened the market; it probably also lowered quality except in the comparatively rare cases where the finest cotton was used. The introduction of man-made fibres in the twentieth century has had the opposite effect. They are cheaper and harder-wearing than natural fibres, crease less readily and dry easily. (Unluckily ease of drying implies low absorption of moisture – hence the clammy feel of some synthetics.) In the kitchen plastic bowls and aluminium saucepans have displaced enamel, being cheaper, lighter and less liable to damage. But the plastic watering-can though cheap and light, has a much shorter life than its old-fashioned competitor made with galvanised iron. The history of the motor car shows no long-term trend towards an ephemeral product. Early cars, heavily built, with a separate chassis, lasted well. Some of their lighter successors built twenty or thirty years ago proved too flimsy to be durable, but in recent years galvanisation has greatly extended the life of the lightly built car. The good reputation earned by careful manufacturers has forced others to conform once more to higher standards.*

ECONOMIES OF TIME AND LABOUR

Economy of materials cheapens goods, widens the market and promotes economic satisfaction unless discontent with poor quality outweighs the pleasures of plenty. The long-term effect is likely to be increased demand for materials and a larger draft on the world's resources. Economies of time and of labour lead to the same result: cheapness, a wider market and a larger consumption of exhaustible resources. In the short term economies of time, for example the railway, or computerised stock-control, reduce the quantity of goods in transit or held as stock, in relation to a given output or turnover. In the longer run the higher levels of efficiency result in higher output. Where economies of labour occur, the impact on output and resources is immediate and direct. Inventions that enable one man to do the work of two or of ten, usually lead at once to higher output: the cotton industry in the eighteenth century is a classic case. Flow (as against batch) methods of production – in paper-making, and in the production of soap, steel, pottery, and cars, to name but a few industries – save both time and labour. They are

* The limited evidence in the author's household by and large suggests that durable consumer goods really do last. A wallet new in 1947 is still in regular use. A Bendix washing machine bought in 1961 went well for fifteen years. A Hotpoint refrigerator bought in 1962 was still working in 1986, unrepaired though battle-scarred. A Swanmaid dishwashing machine was giving good service after 23 years. An Austin Cambridge estate car, 1969 make, did its 10,000 miles a year until succumbing to rust in 1986; but a Simca 1500 made in 1959 had nothing to commend it except good looks. It lasted with difficulty for 11 years. Creda spin driers last well but neither a British nor an Italian washing machine survived more than about five years. Steam-irons – three in seven years – are the most persuasive advocates for the theory of planned obsolescence.

not meant to save natural resources whether exhaustible or reproducible.

ECONOMIC GROWTH AND EXHAUSTIBLE RESOURCES

Economic growth is commonly defined as an increase of real income per head after due allowance for maintaining capital intact. Clearly if a family indulged in good food, expensive clothes and long holidays, but spent nothing on the repair and maintenance of their dwelling they might for a time seem to increase their real consumption – putting off abstinence until the discomforts of their crumbling home exceeded the pleasures of champagne and fur coats. But no economist or accountant would allow that they had experienced a genuine rise in their standard of living during their years of extravagance: rather, it would be said, they had been living on capital, like the spendthrift heir who sold the family silver. The accounts of a country or of the whole world are no different, and those who compile national-income statistics take care to make proper allowances for depreciation of capital and for changes in stocks and in reserves of foreign exchange.

Difficulty arises over the definition of capital. Traditionally it has meant the stock of goods that produces or is capable of producing an income – plant and machinery, houses, the goodwill attaching to brand names in business, money and financial assets. Economists have not had much sympathy with wider definitions of capital that include natural resources. It is true that the Physiocrats and Adam Smith regarded land as the principal source of wealth, but they were writing in an age that lived by agriculture and by industry based on agricultural products. Mineral resources, the readily exhaustible part of the world's landed endowment, were little exploited at the time, and it is unreasonable to look for anxious references to the niggardliness of nature in the *Wealth of Nations.* There is less excuse for economists who have read Malthus on population or Jevons on coal, but it is pretty clear that the warnings of Jevons at least were not taken seriously. With few exceptions leading British economists brushed aside any concern about natural resources, which they treated as given – rather like an axiom in mathematics. J S Mill for example argued that technical improvements in the process of manufacture would decrease costs more than the higher price of exhaustible raw materials would raise them. 'The crude material', he explained, 'generally forms so small a proportion of the total cost'. (Mill, *Principles of political economy*, p 703.) When Marshall defined capital 'from the social [not the individual] point of view' he took care to make his axioms clear: for him capital was:

> all things other than land, which yield income that is generally reckoned as such in common discourse . . . the term land being taken to include all free gifts of nature, such as mines, fisheries etc, which yield income. (Marshall, *Principles of economics*, p 78.)

He recognised that in the long run mines would give diminishing returns, but believed that increase of population pressing on food supplies was more to be feared than the exhaustion of mineral reserves. Keynes is best known for his mordant criticism of the Treaty of Versailles and for his pessimistic analysis of employment prospects in a free market; and like most economists he was aware of the dangers of population pressure. For all that, hope predominated in his thought. In *The General Theory of Employment, Interest and Money* he sought to establish rules for achieving full employment, and in a little known essay 'Economic possibilities for our grandchildren' he took a thoroughly Utopian view of the future. In the absence of war and with a stable population, he argued, real income would increase eightfold in a hundred years and the economic problem – scarcity of resources – would be solved, or at least in sight of solution. His Cambridge colleague A C Pigou, writing on the stationary state, took a more cautious view. While upholding the traditional distinction between capital and natural resources or land, he made an important concession:

> Natural resources – the stores of coal, minerals, oil and so on that Nature has laid up – are also liable to depreciation by the processes that yield income. Plainly here, too, some part of the gross output that comes into being is not real incomings. In so far as this is so our state is not in the strictest sense stationary; for whatever we do we cannot make good the 'wear and tear' that is in this way inflicted on Nature's stores. Our concept however is not damaged seriously provided that we set aside an amount of gross output adequate to provide some form of man-made capital equivalent in 'value' to this wear and tear. (Pigou, *Economics of Stationary States*, p 22.)

It is not clear what 'man-made capital' could replace an exhausted coal or copper mine. Geological exploration might maintain supplies, research provide substitutes or demonstrate ways of doing without, but depletion of some natural resources would continue unless renewable alternatives could be found.

It is perhaps not surprising that depletion of natural resources did not trouble economists in the inter-war years when a glut rather than a scarcity of primary products was a major topic of economic debate. Two important discussions of these problems therefore attracted little attention at the time, though they have been much quoted more recently. The first was an article by the young mathematician-philosopher F P Ramsey, 'A mathematical theory of saving' published in the *Economic Journal* in 1928. In it he discussed, as any fundamental theory of savings requires, the duty of the present to the future. Strictly speaking, he argued, future generations were as entitled to consideration as the present: it followed that 'we do not discount later enjoyments in comparison with earlier ones, a practice which is ethically indefensible and arises merely from the weakness of the imagination'. (*Economic Journal*, p 543.) That was a counsel of perfection given man's love of self; it was also probably an unnecessary restriction since the future and especially the distant future, was and is largely unknowable; and it would be pointless to save for the unknown. Ramsey therefore felt obliged after all to introduce a rate of discount for later enjoyments, as other economic writers had done. The American economist Harold Hotelling encountered similar problems when he considered the economics of exhaustible resources in 1931. Like Ramsey, he treated the subject mathematically, but found it hard to suggest an optimum rate of exploitation, when future prices and future rates of interest (to mention only two of the possible variables) were unknown. The upshot of his study closely resembled the sense of the unhelpful English proverb 'We never know the worth of water till the well is dry'.[2]

Concern about the depletion of natural resources has always been stronger in richly endowed America than in crowded Britain. When W S Jevons published *The Coal Question* in 1865 no school of economists followed up his work. When G P Marsh published *Man and Nature* in 1864, the book established him as the founding father of the conservation movement in America. Conservation has attracted attention there ever since, admittedly without saving the forests or even at times the land itself from destruction. It has been easier to rouse opinion against the evils of pollution and loss of amenity than against the depletion of mineral resources arising from industrial demand for raw materials. Pollution became a major concern in America after publication in 1962 of the eloquent and highly successful *Silent Spring* by Rachel Carson, a marine biologist. The book popularised a great deal of ecological research, most of it undertaken since 1945, and alerted the public to the dangers of indiscriminate chemical warfare against pests and weeds. Ten years later, in 1972, alarming forecasts about mineral resources were given widespread publicity. Again it was not economists but on this occasion natural scientists and demographers who startled the world. Their report *The limits to growth* made sombre reading. With somewhat extravagant assumptions about growth rates the authors showed that within 25 years the world would have used up its known reserves of copper, gold and silver, lead, mercury, natural gas, oil, tin and zinc; within 50 years, aluminium, molybdenum, tungsten and platinum would have disappeared; nickel and cobalt would last a little longer, iron until AD 2065 and coal until AD 2083. Several American and the few British economists who deigned to notice this unwelcome intrusion into their territory reviewed *The limits to growth* in a good-humoured and bantering tone. They pointed to the dangers of uncritical extrapolation, a fault only too obvious to those familiar with Jevons's *The coal question.** Reserves, they explained, were known from exploration and had a habit of keeping pace with consumption. If shortages did appear, they would be reflected in higher prices, which would affect economic behaviour by encouraging economy in use and the search for substitutes. Unlike the scientific authors of the report, the economists had boundless confidence in new technology: energy from solar radiation, the fast-breeder reactor, or nuclear fusion; new materials from the laboratory. Their case was essentially an empirical or historical one: resource problems had been solved in the past by substitution and other shifts in market behaviour, and would be solved in future in much the same way.[3]

Pollution and threats to amenity were taken more seriously. England in 1848, with less than half its present population, already

* According to Jevons, coal output in Britain would have reached 2607 million tons by 1961 if it had continued to grow by 3.5 per cent p.a. for a hundred years (op cit, p 213). Actual output in 1961 was 180 million tons. Like Jevons the authors of *The limits to growth* took 3.5 per cent as their central growth rate.

seemed overcrowded to the younger Mill, and men were in danger, he thought, of losing the delights of solitude and of communion with Nature. Pleasures once reserved to the well-to-do – the Lake District, the National Gallery, Venice – are now widely shared, and too many visitors spoil the view. Economists have not been slow to deplore, sometimes in an elitist frame of mind, the loss of amenity when too many people seek to share a public good that is fixed in quantity and cannot be replaced, still less multiplied. Nothing short of depopulation or impoverishment seems likely to remedy this state of affairs that affluence brings about. For pollution, remedies can be found – in the mixed economy by regulation, or in a free market by taxes. A simple fiscal remedy is to make the polluter pay for the pollution he causes. If a pollution tax is imposed on all users of a stretch of river, they will abate their pollution up to the point where further abatement will cost more than the tax. From that point onward rational firms will rather pay the tax. Only lengthy and carefully observed experiment can show whether taxation and market forces will give better results than the current insistence on standards and best practicable means. The plain man's objection to relying on a pollution tax was well put by Pigou many years ago in a discussion of smoke prevention: 'The general interest requires that these devices should be employed beyond the point at which they "pay".'[4]

If pollution taxes were set at a level that required air and water to be restored to their original purity this objection would not apply. The taxes would reduce profits and wages but firms would simply be put on the same footing as the hotel guest, who expects to occupy a clean and tidy room, and knows that his bill will include a sum for cleaning the room again after he has left. Why treat firms more leniently than hotel guests? Mrs Margaret Thatcher used the metaphor of the full repairing lease to illustrate our duty to leave the world as we find it. It is not clear whether she meant her words to be taken seriously, for the policy of the full repairing lease would almost certainly halt economic growth at a stroke. However this may be, the supporters of a pollution tax become helpless, if not speechless, once non-economic considerations enter the argument, much as Adam Smith was prepared to abandon his formidable case against the Navigation Acts if defence was of more importance than opulence.

The social costs of economic change (pollution and loss of amenity) clearly reduce the true rate of economic growth, and few economists or economic historians would disagree. For it is widely accepted that the quality of life is a real if cloudy concept, and that it has to be allowed for in any attempt to measure (say) the gains and losses arising from industrialisation in early nineteenth-century Britain. But it is not widely accepted that the depletion of natural resources should also be brought into the account. If there were no hope of finding substitutes for exhaustible resources, it would certainly follow that successive drafts upon the world's coal, oil, uranium and natural gas would represent an irreplaceable loss, and that minerals capable of being recycled would have to be used over and over again once the world's virgin resources had been exhausted. In these conditions energy used would count as capital consumed because it would have lost the power to generate further income. Minerals would count as renewable resources, albeit renewable at high cost if dumped on a rubbish tip or swept out to the sea in a state of severe dilution. On these terms the true rate of economic growth would fall well below the rate conventionally recorded. But if exploration and invention could put off indefinitely the day when resources were exhausted, there would be no need to treat the gifts of nature as capital, and the rate of economic growth actually recorded could stand. Most leading British and American economists appear to be willing to back man's ingenuity against the niggardliness of nature. (J E Meade in Britain and N Georgescu-Roegen and perhaps K E Boulding in America have dissented from the prevailing view.) Among those who profess other relevant disciplines deep pessimism reigns.

It seems unlikely that the economists will win the argument in view of the continued growth of population and of incomes per head. Population, it is true, grows fastest in Asia, Africa and Latin America where incomes are small and there is little effective demand for metals and fossil fuels. There, population pressure destroys rainforest and leads to soil erosion: the world loses its land rather than the underlying minerals. In Europe, North America and Japan where most of the world's mineral resources are consumed population growth is slower but has by no means come to an end. This is especially true in North America, which grows by immigration as well as natural increase, and maintains its long tradition of rising standards of living. Even low rates of economic growth can make alarming inroads into the world's reserves of metals and fossil fuels once a high level of extraction has been reached. A growth rate of one per cent per annum quadruples extraction rates in 140 years and multiplies them sixteenfold in 280 years. If demand grows at the higher and less probable rate of 2.5

per cent the consumption of minerals quadruples in only 60 years. Man has been drawing heavily on the world's mineral resources for little more than the past 200 years, a small part of recorded history. In that time shortages have not been hard to overcome, but there is no guarantee that history will repeat itself, that market forces, or even back-stop technology,* will indefinitely stave off an ultimate scarcity of raw materials. If, or when, that day arrives future generations will have little reason to thank the spirit of enterprise that animated industrial man.

It is now more than twenty years since the authors of *The limits to growth* made their alarming predictions. Nothing much has changed in that time. For quite unconnected reasons, rates of economic growth have slowed, abating the rate of depletion of mineral resources. As the economists predicted, proven reserves have kept pace with consumption and shortages are not in sight. In the absence of economic pressures the interests of posterity get little attention: at best there is more willingness to recycle, and in Japan if nowhere else energy-saving is taken seriously. The environmental problem that arouses most anxiety is pollution, which concerns the present more than the future. That the interests of posterity continue to be neglected should occasion little surprise. It is not simply 'want of imagination', as Ramsey asserted, but a preference for self, especially in the face of uncertainty. Love of family is in most men and women stronger than love of mankind in general, and posterity to most people would be of little interest if it did not include their own kin. But there is only a slender chance that a given line will last more than a few generations.** Only half the kings and queens who have reigned in England since the Norman Conquest are direct ancestors of Elizabeth II. The population of

---

* Back-stop technology sets limits to the price of scarce resources. For example, if other sources of fuel become expensive, solar energy (a back-stop technology) will set bounds to the rise in fuel costs. Unfortunately, back-stop technology is an economic theory, devised by W D Nordhaus, not an engineering reality.

** In any case, distant ancestors contribute little to the make-up of their remote descendants. In ten generations from now, unless there has been intermarriage, a child will share only one part of his genetic inheritance in 1024 (two to the tenth) with each of his late-twentieth-century ancestors. A kind of earthly immortality is assured to all of us because the natural world is a giant system of recycling. But only an ardent conservationist will look forward with much pleasure to rebirth, perhaps simultaneously, in a worm, a pig, a reed, and an oak.

England in 1066 is estimated at between one and two million. It is quite possible given high mortality rates in childhood, failure to marry, and childlessness in marriage, that only five or ten per cent of those who were then in the 15–45 age-group have descendants alive today. At best the proportion is unlikely to be more than a quarter. If love of family is, after love of self, the most powerful motive for thrift and concern for posterity, the disregard of future generations is understandable.

Environmentalists propose drastic measures to avert the dangers that they foresee: a stop to economic progress and to population growth. Even if these ends were achievable, slow attrition of the world's wealth would continue since exhaustible resources would still be used up, though at a steady rather than an increasing rate. The truly stationary state, as Pigou realised, would not touch non-renewable assets unless it could replace them with some equivalent. In all probability, despite the pressure of environmentalists, there will be no sharp break with our prodigal traditions. So long as growth is less painful than restraint, growth is likely to go on. The economic system will respond to scarcity in due course, but much more slowly than environmentalists would wish. Mankind faces much more urgent problems: unemployment, hunger and the risk of war are the most obvious. The threats to freedom are scarcely less acute, and come from many quarters: corrupt and oppressive government; irresponsible ownership of the mass media; religious fanaticism; the arrogance of experts. Among these woes the dangers arising from economic growth seem to most men distant if not imaginary. More than a century of slowly rising real income has raised expectations, and the poor may well feel cheated if they are now told that growth has got to stop. It is fairly easy for men who hold a comfortable station in life to recommend abstinence to others. Adam Smith as consumer had expensive taste in clothes, but as moral philosopher he regarded the pursuit of wealth as in itself 'in the highest degree contemptible and trifling.' As economist, he acknowledged that money-making was useful if it led men to satisfy the wants of those less fortunate or less enterprising than themselves. Some modern economists have suggested a curbing of economic appetites, a retreat to the simple life and to 'wantlessness' as the proper response to the niggardliness of nature.[5] But how many, even among environmentalists, are ready to follow this hard advice if it interferes with their material comforts? Pretty certainly, the average man will not be quick to tighten his belt for the sake of future generations, or out of fear of what they may think of him.

When long years hence the fires burn low,
Who will praise our spendthrift ways
Or fondly dwell on Arkwright's show?

REFERENCES

1. H J Hopkins, *A span of bridges*, (Newton Abbot, 1970), p 33; A Burstall, *History of mechanical engineering* (1965), p 213; Manchester Steam Users' Association, *A sketch . . . of the past fifty years' activity* (1907), p 7; s.c. on the North British Railway (Tay Bridge) Bill (PP 1880 XII), qq 617–32, 1218, 2486; *Report of the court of enquiry into the Tay Bridge disaster* (PP 1880 XXXIX), pp 6–10, 12–14, 28, 31–7, 41; *Evidence* (ibid.), qq 16615, 16764, 16831 ff, 16915 ff; App. 2 (ibid.), p ii; *Report on the loss of the steamship* Titanic (PP 1912–13 LXXVI), pp 16–17, 24–30, 36.
2. Rates of depreciation from the annual reports of British Leyland, British Telecom, Rolls Royce and Amstrad; B W Clapp, *John Owens Manchester merchant* (Manchester, 1965), p 82; *The Observer*, 28 April 1991; A E Higginson, 'Disposal of refuse by incineration' (NSCA, *Clean air conference Southport 1970*), p 30; J S Mill, *Principles of political economy* (repr 1923), ed Ashley, pp 702–3; J E Cairnes, *Some leading principles of political economy* (1874), pp 132–5; Marshall, *Principles of economics*, pp 78, 167, 179–80; J M Keynes, *Essays in persuasion* (1933 edn), pp 358–73; A C Pigou, *Economics of stationary states* (1935), p 22; R F Harrod, *Life of J M Keynes* (1951), pp 394, 399; *Economic Journal* (1928), pp 543–59; *Journal of Political Economy* (April 1931); F P Wilson ed, *The Oxford dictionary of English proverbs* (Oxford, 1970), p 435.
3. Dennis L Meadows *et al.*, *The limits to growth*, (1972), pp 56–60; W D Nordhaus, 'World dynamics: measurement without data' (*Economic Journal*, December 1973), pp 1156–83; S Gordon, 'Today's apocalypses and yesterday's' (*American Economic Review*, May 1973), pp 106–10; N Rosenberg, 'Innovative response to materials shortage', ibid., pp 111–18; R M Solow, 'The economics of resources or the resources of economics', ibid. May 1974, pp 1–14; T B Cochran, *The liquid metal fast-breeder reactor* (Baltimore, 1974); J Stiglitz, 'Growth with exhaustible natural resources: efficient and optimal growth paths' (*Review of Economic Studies*: symposium on the economics of exhaustible resources, 1974), pp 123–37; H P Chao, *Economies with exhaustible resources* (New York, 1979); W Beckerman, *In defence of economic growth* (1974).
4. Mill, *Principles*, p 750; E J Mishan, *The costs of economic growth* (1967); F Hirsch, *The social limits to growth* (1976); W Beckerman, *In defence of economic growth*, esp. Chapter 6; A C Pigou, *Wealth and welfare* (1912), p 159 n.

5. Adam Smith, *The theory of moral sentiments* (6th edn 1790), I, 463–7; E F Schumacher, *Small is beautiful* (1973); K E Boulding, *A reconstruction of economics,* (New York, 1950), pp ix, 27, *Collected papers* II ' What is economic progress?' (1961), p 177; J K Mehta, *Rhyme, rhythm and truth in economics* (1967), p 27.

Index